16th Symposium of Structural Dynamics (DYNKON 2019)

MATEC Web of Conferences Volume 285 (2019)

Kombornia, Poland
22 - 24 May 2019

Editors:

Tadeusz Burczynski
Bartosz Miller
Leonard Ziemianski

ISBN: 978-1-5108-9070-1

Printed from e-media with permission by:

Curran Associates, Inc.
57 Morehouse Lane
Red Hook, NY 12571

Some format issues inherent in the e-media version may also appear in this print version.

No changes have been made to the content of these proceedings. There may be changes to pagination and minor adjustments for aesthetics.

Printed by Curran Associates, Inc. (2019)

For additional information, please contact EDP Sciences – Web of Conferences at the address below.

EDP Sciences – Web of Conferences
17, Avenue du Hoggar
Parc d'Activité de Courtabœuf
BP 112
F-91944 Les Ulis Cedex A
France

Phone: +33 (0) 1 69 18 75 75
Fax: +33 (0) 1 69 28 84 91

contact-edps@webofconferences.org

Additional copies of this publication are available from:

Curran Associates, Inc.
57 Morehouse Lane
Red Hook, NY 12571 USA
Phone: 845-758-0400
Fax: 845-758-2633
Email: curran@proceedings.com
Web: www.proceedings.com

TABLE OF CONTENTS

DYNAMIC PROPERTIES OF FAÇADE FRAME SCAFFOLDINGS ... 1
Bec Jaroslaw, Blazik-Borowa Ewa

COMPARISON OF THE VALUE OF PASSIVE SAFETY COEFFICIENTS FOR CAR CRASH WITH LIGHTING COLUMN MADE OF ALUMINUM OR COMPOSITE MATERIAL BASED ON NUMERICAL SIMULATION ... 9
Danek Wojciech, Gasiorek Damian

THE GAO BEAM UNDER A MOVING INERTIAL LOAD AND HARMONIC COMPRESSION ... 15
Dyniewicz Bartlomiej, Bajer Czeslaw

DYNAMIC STRESS INTENSITY FACTORS OF INTERACTING BRANCHED CRACKS ... 23
Fedelinski Piotr

APPLICATION OF THE TWO-PARAMETER SINGLE SHOOTING METHOD FOR NATURAL VIBRATION ANALYSIS OF EULER-BERNOULLI BEAMS ... 31
Holubowski Ryszard

HISTORICAL CHURCH DIAGNOSTIC WITH REGARDS TO IMPACT OF TRAFFIC INCLUDED VIBRATIONS ... 37
Koziol Krzysztof

REDUCTION OF VIBRATION IN RAILWAY SURFACES ... 43
Koziol Krzysztof

OPTIMAL POSITIONS OF TRANSLATIONAL VIBRATION ABSORBERS IN VIBROISOLATION OF MECHANICAL PRESS ... 51
Latas Waldemar

FREE VIBRATION OF FRAME STRUCTURES MADE OF ZENER TYPE VISCOELASTIC MATERIAL ... 59
Lewandowski Roman, Wielentejczyk Przemyslaw

EXPERIMENTAL VERIFICATION OF THE NUMERICAL MODEL OF A REINFORCED CONCRETE ARCH ... 67
Markiewicz Barbara, Piatkowski Grzegorz, Szyszka Lukasz, Ziaja Dominika

DYNAMIC PROBLEMS IN THE OPERATION OF OVERRESONANT VIBRATING SCREENS ... 75
Matachowski Filip, Badura Slawomir

FREQUENCY OPTIMISATION OF COMPOSITE CYLINDER USING AN EVOLUTIONARY ALGORITHM AND NEURAL NETWORKS ... 83
Miller Bartosz, Ziemianski Leonard

ARRIVAL TRAFFIC SEPARATIONS MODELLING BY USING CONTINUOUS PROBABILITY DISTRIBUTIONS ... 89
Pawelek Adrian, Lichota Piotr

NUMERICAL MODEL OF TRACKED VEHICLE CROSSING INLAND WATER ... 97
Pawlak Mariusz, Slawski Sebastian, Duda Slawomir

DAMPING PARAMETERS OF A STEEL STRUCTURE WITH A VISCOELASTIC LAYER IN THE BASE NODE ... 105
Pawlak Zdzislaw, Zwolinski Dawid

DYNAMIC PERFORMANCE OF A VEHICULAR BRIDGE WITH LIGHTWEIGHT FRP COMPOSITE STRUCTURAL ELEMENTS ... 113
Siwowski Tomasz, Rajchel Mateusz

RESEARCH OF INFLUENCE OF VIBRATION IMPACTS OF CHIMES ON THE STEEPLES OF ST. ISAAC\'S CATHEDRAL (SAINT PETERSBURG) AFTER RECONSTRUCTION THEREOF ... 121
Smirnova Elena E., Savin Sergey N., Larin Denis V.

ON DYNAMIC INTERACTION BETWEEN MECHANICAL SYSTEMS AND SELECTED ELECTRIC MOTORS ... 129
Szolc Tomasz, Konowrocki Robert, Pochanke Andrzej

VIBRATION DAMPING IN ROAD BRIDGES – TWO CASE STUDIES OF FULL SCALE EXPERIMENTS ... 137
Szynal Daniel, Janas Lucjan

STABILITY OF A HYDRAULIC TELESCOPIC CYLINDER SUBJECTED TO GENERALISED LOAD BY A FORCE DIRECTED TOWARDS THE POSITIVE POLE ... 145
Uzny Sebastian, Kutrowski Lukasz

DYNAMIC TESTS OF COMPOSITE FOOTBRIDGE SEGMENT – EXPERIMENTAL AND NUMERICAL STUDIES 153
Wiczenbach Tomasz, Ferenc Tomasz, Pyrzowski Lukasz, Chróscielewski Jacek
ON ESTIMATION OF OCCUPANT SAFETY IN VEHICULAR CRASHES INTO ROADSIDE OBSTACLES USING NON-LINEAR DYNAMIC ANALYSIS 161
Wilde Krzysztof, Tilsen Arkadiusz, Burzynski Stanislaw, Witkowski Wojciech
PARAMETRIC VIBRATIONS IN MECHANICAL SYSTEMS ON THE EXAMPLE OF A MINING HOIST INSTALLATION 169
Wolny Stanisław
THE EFFECTS OF IRREGULARITIES AND MISALIGNMENTS OF THE GUIDE STRINGS ON STRESS LEVELS IN SELECTED LOAD-BEARING ELEMENTS OF A CONVEYANCE 177
Wolny Stanisław, Drzewosz Agata
ANOMALY DETECTION IN THE CONCRETE ARC GIRDER SUBJECTED TO FATIGUE TEST 185
Ziaja Dominika, Nazarko Piotr
Author Index

Statement of Peer review

In submitting conference proceedings to *Web of Conferences*, the editors of the proceedings certify to the Publisher that

1. They adhere to its **Policy on Publishing Integrity** in order to safeguard good scientific practice in publishing.
2. All articles have been subjected to peer review administered by the proceedings editors.
3. Reviews have been conducted by expert referees, who have been requested to provide unbiased and constructive comments aimed, whenever possible, at improving the work.
4. Proceedings editors have taken all reasonable steps to ensure the quality of the materials they publish and their decision to accept or reject a paper for publication has been based only on the merits of the work and the relevance to the journal.

Title, date and place of the conference

DYNKON 2019, 16th Symposium of Structural Dynamics
22-24.05.2019
Kombornia, Poland

Proceedings editor(s):

Tadeusz Burczyński
Bartosz Miller
Leonard Ziemiański

Date and editor's signature

24.05.2019

17, avenue du Hoggar - PA de Courtabœuf - BP 112 - 91944 Les Ulis Cedex A (France)
Tél. : 33 (0)1 69 18 75 75 - Fax : 33(0)1 69 07 45 17 - www.edpsciences.org

MATEC Web of Conferences **285**, 00001 (2019) https://doi.org/10.1051/matecconf/201928500001
DYNKON 2019

Dynamic properties of façade frame scaffoldings

Jarosław Bęc[1,*], and *Ewa* Błazik-Borowa[1]

[1]Lublin University of Technology, Faculty of Civil Engineering and Architecture, Nadbystrzycka 40, 20-618 Lublin, Poland

Abstract. The paper presents data on natural frequencies and damping parameters of scaffoldings. The analyzes were made mainly on the basis of acceleration measurements made on scaffoldings on construction sites. These studies were supplemented by numerical analyzes, which allowed to show changing of the natural frequencies of the scaffoldings as the effect of the structure exploitation. Finally, it was found that the scaffoldings are structures with low natural frequencies ranging from 1.0 Hz to 4.0 Hz and a logarithmic decrement of damping at 0.06 in most cases, however differences can be observed with the values as low as 0.03.

1 Introduction

Scaffoldings are rod structures, stiffened with use of decks. Principles of operation of scaffoldings as a frame system under the influence of static loads have been described in [1]. In this work, it was also shown that dynamic actions are a significant load for scaffolding. Performing dynamic analyzes requires examination of the dynamic properties of the scaffolding, in particular the damping parameters, the natural frequencies and the accompanying mode shapes of vibrations. In this paper there were used the results of research of 120 scaffoldings, which was carried out in the years 2016-2018 as a part of the PBS3/A2/19/2015 project "Modelling of Risk Assessment of Construction Disasters, Accidents, and Dangerous Incidents at Workplaces Using Scaffoldings" ORKWIZ, financed by the Polish National Center for Research and Development. As part of this project, among others, such tests as scaffolding and damage inventory (comp. [2], [3]), measurements of environmental parameters (comp. [4], [5]), loads measurements (comp. [1], [6], [7]), parameters of the work environment (comp. [8]) and measurements of accelerations of scaffolding vibrations (comp [1], [9]) were made. Based on measurements of scaffolding accelerations, frequencies of free vibrations were determined. For selected scaffoldings, damping parameters were calculated, and the problem of users' mass impact on the dynamic behavior of scaffolding was discussed. The methodology of the research is presented here on the example of scaffolding, marked with the symbol W09 in the ORKWIZ project.

* Corresponding author: j.bec@pollub.pl

2. Description of research methods

2.1. Measurements of accelerations of scaffolding vibrations

Measurements of vibrations accelerations in selected points of scaffolding were made using Brüel & Kjær measuring system. In most tests two uniaxial accelerometers and two triaxial accelerometers were used. Accelerometers were placed on top-level frames. The arrangement of the accelerometers is presented in Fig. 1 and the accelerometer mounted on the scaffolding frame is shown in Fig. 2a. The excitation was generated by one person, causing horizontal oscillations along the scaffolding, in the direction perpendicular to the scaffolding plane and bouncing vertically. The measurement frequency was 8192 Hz. The excitations were implemented below the highest level of decks in three locations, also shown in Fig.1.

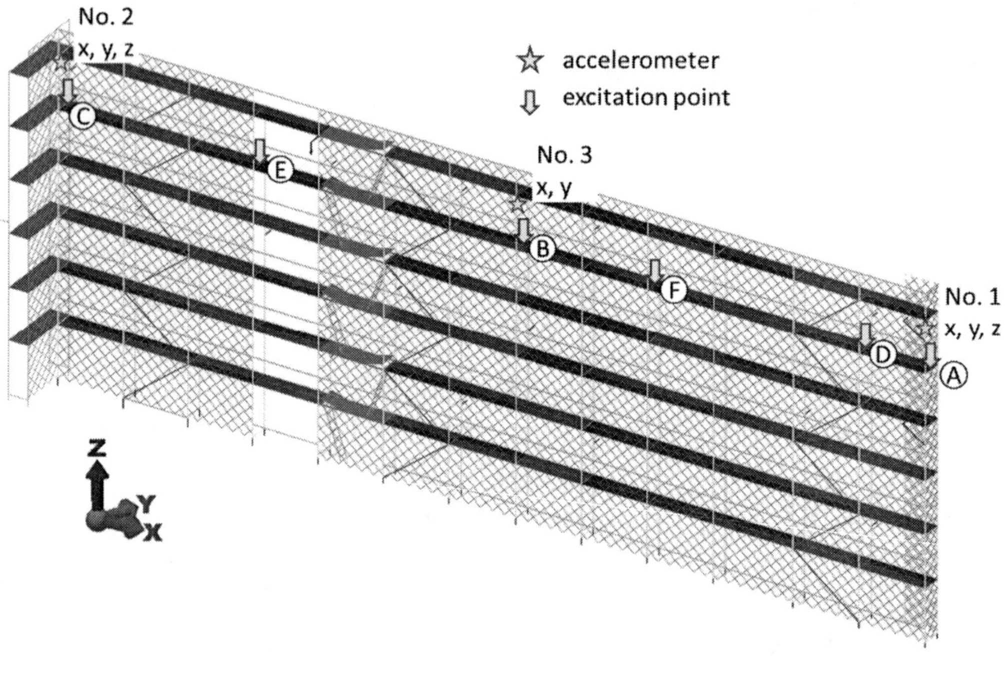

Fig.1. Scheme of the scaffolding with the locations of measurement and excitation points

2.2. Analysis of measured signals

The frequencies of free vibrations and scaffolding damping parameters were determined based on the measured time series of accelerations. For each of the measured signals, an FFT (Fast Fourier Transform) analysis was performed. Based on these results, the frequencies of free vibrations were determined.

The determination of damping parameters was limited to the determination of the logarithmic decrement of damping. The signals were filtered using Butterworth digital filter of the order 8. In the case of the results shown in Fig. 2b, a filter with a bandwidth of 1.5 Hz to 2.5 Hz was used and the analyzed frequency of free vibrations was $f_s = 2$ Hz. As a result of the filtering, the waveforms shown in Fig. 2c were obtained. A fragment of vibrations with damping was selected from the filtered time series and local maxima were determined for this fragment. Then the points locations were approximated by a curve with the equation:

$$y_a = A e^{-\gamma t}, \tag{1}$$

where: γ – damping coefficient.

Logarithmic decrement of damping is calculated with the formula:

$$\Delta = \frac{\gamma}{\omega_s}, \tag{2}$$

where: $\omega_s = 2\pi f_s$ – circular frequency of free vibrations, γ – damping coefficient.

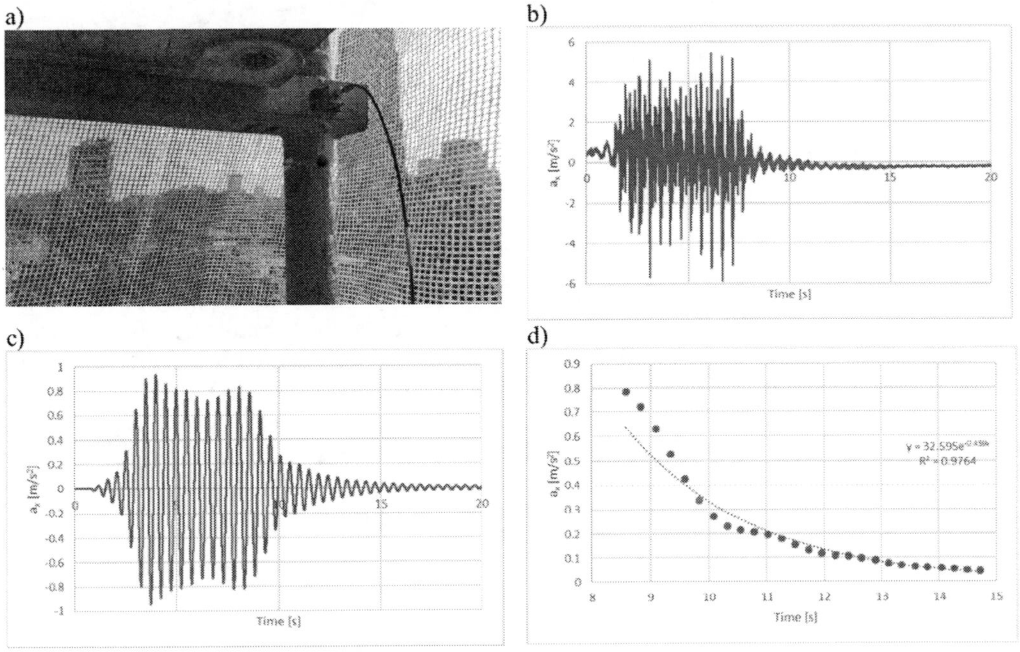

Fig.2. Determination of the damping coefficient: a) accelerometer mounted at the scaffolding frame b) fragment of the time series, c) fragment of the time series after filtration, d) curve fitted to the maxima of accelerations

2.3. Numerical model of the scaffolding

The last method of research presented here is numerical analysis of W09 scaffolding. The model considers the state of the structure and the effects of the scaffolding exploitation, for example the loosening of anchors in the wall ([1], [10]). This was achieved by calibrating the numerical model based on measurements of the free vibrations' frequencies. Figure 3 presents the model. The structure of the scaffolding was modeled with beam elements, while the decks – with shell elements. Connections between handrails and stands, braces and stands are modeled as hinge joints. The other connections are stiff. Supports, modeling the foundation on the ground and anchors in the wall, are modeled using hinged supports.

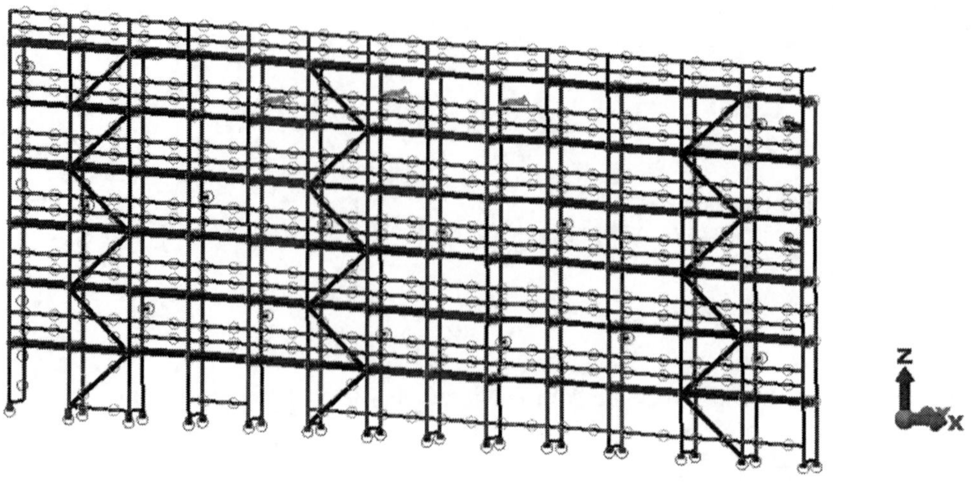

Fig. 3. Static scheme of the scaffolding

3. Analysis of the results

3.1. Frequencies of free vibrations and natural frequencies

For 120 scaffoldings analyzed in the ORKWIZ project, numerical models were made. The models were verified on the basis of acceleration measurements. Next, dynamic analyzes were performed for each of the scaffoldings, which resulted in natural frequencies and mode shapes. Fig. 4 shows, how the values of the first and second scaffolding natural frequency is distributed, without taking into account the additional mass of users. The first mode shape for 100 scaffoldings was observed as vibrations along the scaffolding plane. For 40 of the scaffoldings the second mode shape was also longitudinal. Fig. 5 shows the first four mode shapes of the W09 scaffolding.

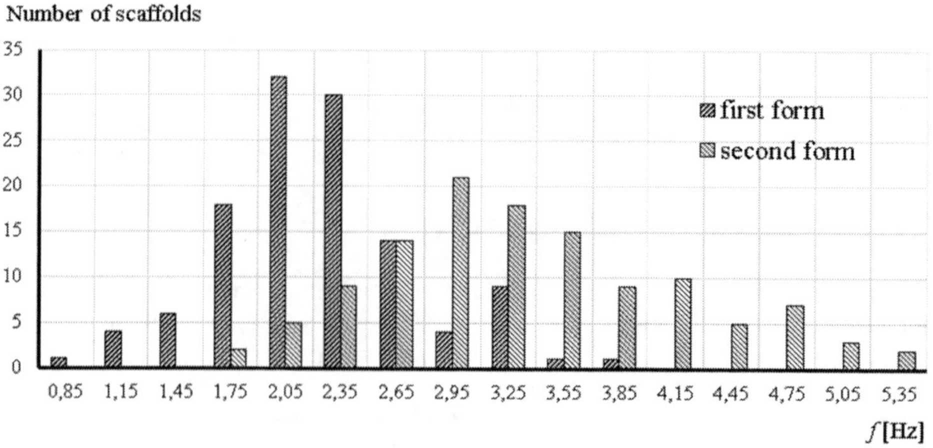

Fig. 4. The distribution of the values of the first two natural frequencies

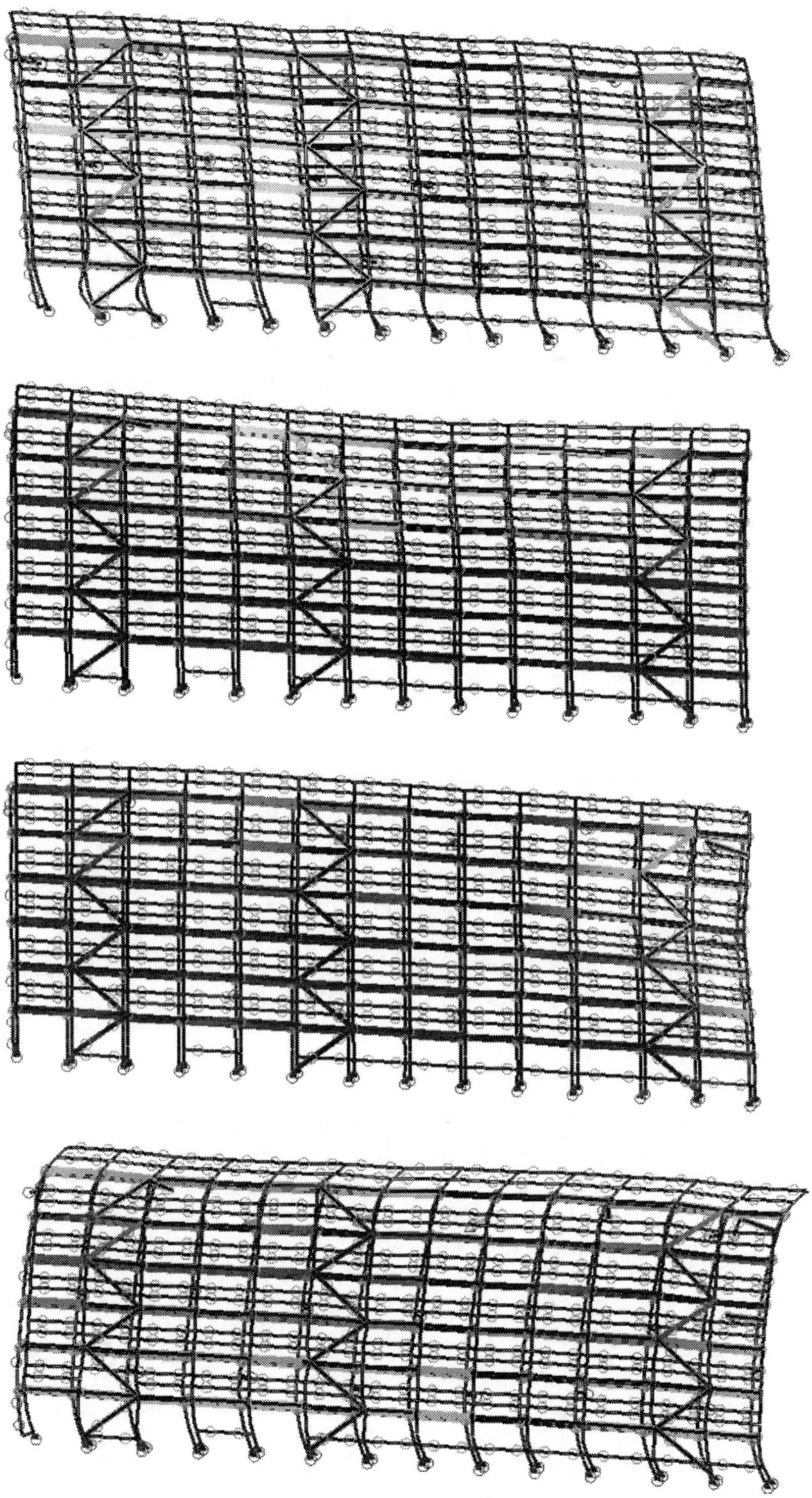

Fig. 5. Mode shapes of the W09 scaffolding: a) f_1=2.3 Hz, b) f_2=4.18 Hz, c) f_3=4.54 Hz, d) f_4=5.04 Hz.

The weight of scaffolding users can affect the frequency of vibrations, which will be shown on the example of W09 scaffolding. The scaffolding mass is 5.9 tons. The permissible load according to EN 12811-1 [11] according to class 3 is 200 kg/m^2 on one deck level. The total additional mass in case of W09 scaffolding is 4.6 tons, that is 75% of the scaffolding's mass. Fig. 6 shows how the first four natural frequencies of the W09 scaffolding change with the increase of mass, located at the highest level of decks. Including the mass of users for each scaffolding will result in reduction of the frequency values. However, this change depends on the design of the structure, and in particular the layout of the anchors and braces.

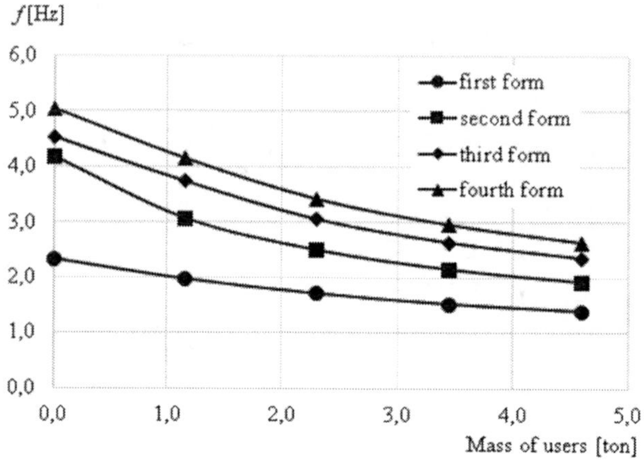

Fig. 6. The influence of users' mass on the natural frequencies of the W09 scaffolding.

3.2. Damping parameters

Table 1 summarizes the results of the logarithmic decrement of damping for 30 signals measured on the W09 scaffold. The mean value and standard deviation of the logarithmic decrement of damping are 0.0292 and 0.0088, respectively. The values determined for several scaffoldings are listed in Table 2. The scaffold symbols used in the table are symbols used in the ORKWIZ project (see [1]).

Table 1. Determination of damping parameters for scaffolding W09

Location and direction of excitation	Measurement point and direction	Damping coefficient γ [rad/s]	Frequency of free vibrations f_s [Hz]	Logarithmic decrement of damping
Ax	1x	0.246	2.13	0.0184
Ax	1x	0.228	2.13	0.0171
Dx	1x	0.346	2.06	0.0267
Dx	1x	0.328	2.06	0.0253
Bx	1x	0.436	1.94	0.0358
Ex	1x	0.459	2	0.0365
Ex	1x	0.634	2	0.0505

Fx	1x	0.411	2	0.0327
Ax	2x	0.244	2.13	0.0183
Ax	2x	0.226	2.13	0.0169
Dx	2x	0.345	2.06	0.0266
Dx	2x	0.308	2.06	0.0238
Bx	2x	0.406	2	0.0323
Cx	2x	0.414	2.06	0.0319
Cx	2x	0.432	2	0.0344
Ex	2x	0.302	2	0.024
Ex	2x	0.541	2	0.0431
Fx	2x	0.41	2	0.0326
Fx	2x	0.351	1.94	0.0288
Ax	3x	0.234	2.13	0.0175
Ax	3x	0.232	2.13	0.0174
Dx	3x	0.283	2.06	0.0218
Dx	3x	0.318	2.06	0.0245
Bx	3x	0.434	1.88	0.0368
Cx	3x	0.34	2.06	0.0262
Cx	3x	0.428	2	0.0341
Ex	3x	0.42	2	0.0334
Ex	3x	0.598	2	0.0476
Fx	3x	0.408	2	0.0325
Fx	3x	0.348	1.94	0.0286

Table. 2. Logarithmic decrement of damping for selected scaffoldings

Scaffolding symbol	**First natural frequency [Hz]**	**Logarithmic decrement of damping**	
		Mean value	**Standard deviation**
W09	2.04	0.0292	0.0088
P04	3.11	0.0632	0.0133
P05	3.08	0.0540	0.0156

P06	3.50	0.0571	0.0146
P10	1.83	0.0553	0.0119
E05	3.10	0.0617	0.0146

4. Conclusions

The first natural frequencies of scaffoldings without users are less than 4 Hz. The first mode shape is usually vibrations along the scaffolding plane, while the second mode shape is in most cases vibrations in the perpendicular horizontal direction.

Due to the low mass of scaffolding structures, the natural frequency of scaffolding with the mass of users included is even lower than the frequencies given in Figure 4. On the other hand, low natural frequencies cause the scaffolds to be susceptible to wind induced vibrations and vibrations generated by the movement of the users at the scaffolding. In the ORKWIZ project, it was found that the movement of the structure is perceptible to the users while walking on the scaffolding, and in work [9] the increase in scaffolding structure effort due to dynamic interactions was confirmed.

Performing dynamic analyzes of scaffolding requires knowledge of damping parameters. For most of the analyzed scaffoldings it can be assumed that the logarithmic decrement of damping for scaffolding is about 0.06. However, for the scaffolding W09 presented here in this paper the calculated value was about 0.03. In the future it is necessary to determine the damping parameters with more advanced methods. In addition, damping in two directions of vibrations should be investigated.

References

1. E. Błazik-Borowa, *Load and Actions on Scaffoldings as Engineering structures* (in Polish) (Lublin University of Technology, Lublin, 2018)
2. M. Pieńko, A. Robak, E. Błazik-Borowa, J. Szer, International Journal of Civil, Environmental, Structural, Construction and Architectural Engineering, **12**, 2 (2018)
3. Ł. Borowski, M. Pieńko, P. Wielgos, BGC Geomatics, 8071470 (2017)
4. I. Szer, E. Błazik-Borowa, J. Szer, Arch. Civ. Eng. **3**, 63 (2017)
5. M. Jabłoński, I. Szer, J. Szer, JCEM, **24**, 6 (2018)
6. T. Lipecki, P. Jamińska-Gadomska, E. Błazik-Borowa, AIP Conf. Proc., **1922** (2018)
7. T. Lipecki, P. Jamińska-Gadomska, A. Sumorek, E. Błazik-Borowa, J. Bęc, JPCS **1101** (2018)
8. H. Bojar, F. Silveira, M. Rebelo, E. Czarnocka, K. Czarnocki, Annals of Agricultural and Environmental Medicine, 100534 (2018)
9. P. Cyniak, E. Błazik-Borowa, J. Szer, T. Lipecki, I. Szer, AIP Conf. Proc., **1922** (2018)
10. P. Jamińska-Gadomska, J. Bęc, T. Lipecki, A. Robak, Arch. Civ. Eng., **64**, 1 (2018)
11. EN 12811-1. Temporary structures applied at construction sites. Part 1: Scaffolds. Conditions for development and general principles for designing

Comparison of the value of passive safety coefficients for car crash with lighting column made of aluminum or composite material based on numerical simulation

Wojciech Danek[1,*], and *Damian Gąsiorek*[2]

[1]Silesian University of Technology, Institute of Theoretical and Applied Mechanics, Konarskiego 18A, Gliwice, Poland
[2]Silesian University of Technology, Institute of Theoretical and Applied Mechanics, Konarskiego 18A, Gliwice, Poland

Abstract. Numerical simulations have become more and more important in recent years for the design of supporting road equipment structures. These structures are important for ensuring the proper level of passive safety for drivers and passengers. These structures also allow the reduction of costs of entry for new products on the market and reduction of the total cost of production. In this paper was shown comparison of passive safety coefficient ASI and THIV for lighting column made of composite material and aluminum based on experimentally verified numerical simulation. FEM was used to carry out the simulation which was implemented in LS Dyna to determine passive safety coefficients, whose algorithm was written in MATLAB. Authors presented parameters of numerical model of the car and lighting column with the ground. The accepted boundary and initial conditions were chosen in order to show the behaviour of the lighting column during a collision with a car.

1 Introduction

A modern trend for design elements included in road infrastructure is assurance of maximum safety for drivers and passengers. For this purpose, each newly designed infrastructural element is subject to a series of numerical analyses whose purposes are to determine properties of the element during the impact by the vehicle. As a result, the effect of the passive safety of passengers is determined. The results of the numerical simulations are verified by experimental studies. Based on the data presented in a report by the Polish police in 2017, there were nearly 33,000 traffic accidents, in which 3,000 people were killed. A significant proportion of these accidents were collisions with all sorts of obstacles in the forms of trees and lighting columns. In order to increase the passive safety of passengers, in the case of lighting columns, various efforts have been used to significantly reduce strength overload, which arises at the moment of impact, below a level dangerous to

* Corresponding author: wojciech.danek@polsl.pl

humans. research related to this issue was the subject of earlier work carried out by the author [1-2]. Issues related to the research regarding lighting columns are the subject of numerous scientific publications. These publications encompass experiments and virtual tests carried out in various computing environments [3-5]

Supporting road equipment structures like lighting columns can, during a collision, constitute a threat to drivers and passengers. Therefore, when this element of road equipment is designed, one should seek to increase passive safety for drivers and passengers. For this purpose, European standard EN 12767 distinguishes three categories of passive safety [6]:

- High energy absorbing (HE),
- Low energy absorbing (LE),
- Non-energy absorbing (NE).

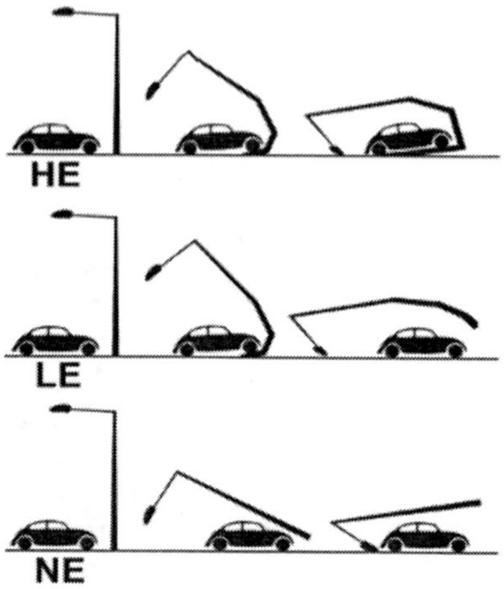

Fig. 1. Level of energy absorption according to EN 12767.

Constructions qualified as HE class are intended to significantly reduce the car's velocity and to reduce the risk of crashes with another car or with other elements of road infrastructure. Lighting columns qualify within two classes are intended to reduce the car's velocity with low risk of significant damage to health.

In order to determine to which class of passive safety a particular construction will be qualified, there are two types of coefficients: Acceleration Severity Index (ASI) and Theoretical Head Impact Velocity (THIV). ASI allows determination of a nuisance motion of the vehicle for people sitting near selected measuring points (usually a center of gravity). It is considered a determinant of accident severity to passengers in the vehicle upon impact with an obstacle. It is determined from the dependence shown below [7-9]:

$$ASI(t) = \left[(\bar{a}x / \hat{a}x)^2 + (\bar{a}y / \hat{a}y)^2 + (\bar{a}z / \hat{a}z)^2\right]^{1/2} \tag{1}$$

Where:

$\bar{a}_x$, $\bar{a}_y$, $\bar{a}_z$ – components of acceleration of the chosen point P of the vehicle from the movable averaged time interval $\delta = 50$ms,

$\hat{a}_x$, $\hat{a}_y$, $\hat{a}_z$ – limit value of acceleration along X, Y, Z axes of the vehicle.

For passengers using seat belts, limit values of acceleration are equal to:

- $\hat{a}_x = 12$g,
- $\hat{a}_y = 9$g,
- $\hat{a}_z = 10$g,

Theoretical Head Impact Velocity is the velocity of the head during an impact the surface inside the vehicle and is determined from the dependence shown below [7-9]:

$$THIV = [V_x^2(T) + V_y^2(T)]^{1/2} \tag{2}$$

Where:

$V_x(T)$, $V_y(T)$ – velocity of theoretical head at the time of impact in the surface inside the vehicle.

Values of these two coefficients are determined for different velocities depending on the road where this obstacle is located.

2 Numerical model

Numerical simulations were performed using a numerical model of Suzuki Swift which met requirements related to mass of the vehicle and center of gravity contained in the standard EN 12767 [6]. Due to the complicated construction of the vehicle and incomplete information about the parameters of the elements included in the vehicle, a model made by the National Crash Analysis Center (NCAC) was used in the simulation. The data relating the parameters of mass and geometry to the elements that have been used for the discretization of the model are shown in Table 1.

Table 1. Geometric parameters of the vehicle and division on finite elements

Vehicle weight [kg]	860
Wheel base [mm]	2371
Distance the center of mass from the front axle [mm]	1128
Number of nodes	19271
Number of solid elements	820
Number of beam elements	4
Number of shell elements	15330
Total number of finite elements	16154
Number of parts	242
Number of material models	6

The lighting column with its inspection door were modelled as shell elements for which the wall thickness was set as specified by the manufacturer. The data relating the parameters of mass and geometry to the elements that have been used for the discretization of the model are shown in Table 2. Presented data are in accordance with standards [10-12]

Table 2. Geometric parameters of the lighting column and division on finite elements

Height of the lighting column [m]	7
Thickness of the wall of the lighting column [mm]	4
Net Weight column [kg]	
Composite	40
aluminum	65
Dimensions of the ground	According to Norm EN 12767
Number of nodes	593897
Number of solid elements	430775
Number of shell elements	87121
Number of material models	4

In the case of modeling the lighting column, it was decided to include ground for the model, whose dimensions were determined from the European standard EN 12767. It was decided to model the assembly of the lighting column this way in order to reflect the real conditions at the time of impact with a vehicle.

As a model for the material of the lighting column made of aluminum "Material Model 24: Piecewise Linear Isotropic Plasticity" while for construction made of composite material "Material Model 054 Enhanced Composite Damage" were chosen. Parameters for the composite material and aluminum was determined based on experimental research made on MTS Criterion Model 43 was chosen. Parameters for ground material model was taken from literature [13]

As a model for the material of the ground "Material Model 05: Soil And Foam"

Additional conditions related to displacements of individual elements and velocity were imposed on the system. In the case of conditions related to displacements, all degrees of freedom were removed in the nodes located on the outer walls of the ground. Linear velocity equal 35 and 50km/h [7-9] was determined for all nodes of the car and angular velocity was determined for wheels. Additionally, in order for the conditions to be as close to reality as possible during the simulation, the gravitational force acted in the system.

3 Results of numerical simulations

In this chapter results of numerical simulation was shown. Numerical simulation model was verified by experiment for car crash with composite lighting column, where value of r-Person correlation coefficient was equal 0,724. Results of Passive Safety coefficient ASI for aluminum and composite material were shown on Fig. 2.

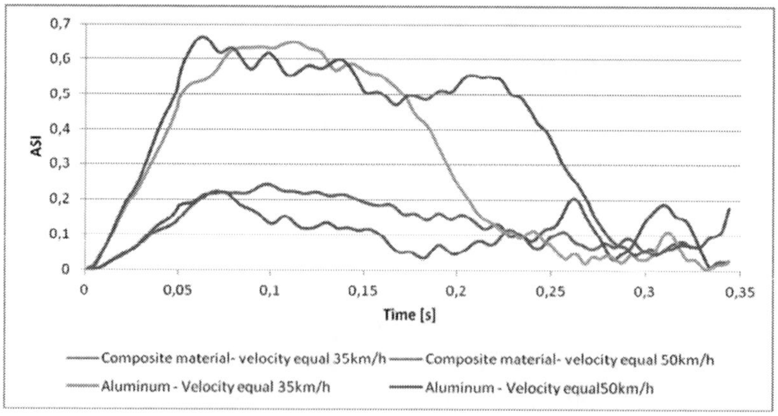

Fig. 2. Values of ASI coefficient for aluminum and composite material lighting column during crash.

Second Passive Safety coefficient THIV for lighting column was equal:

- For Composite material – velocity equal 35km/h - 32,1km/h
- For Composite material – velocity equal 50km/h - 39,6km/h
- For Aluminum material – velocity equal 35km/h – 24,6km/h
- For Aluminum material – velocity equal 50km/h - 29,2km/h

Based on result obtained from numerical simulation can be observed that construction made of aluminium are characterized by higher value of passive safety coefficient ASI. Value of this coefficient for lighting column made of aluminium is almost three times higher than for lighting column made of composite material. In case of second passive safety coefficient THIV can be observed that construction made of aluminum are characterized by lower value of this coefficient than construction made of composite material. Qualitive analysis for every presented simulations was shown on Fig. 3.

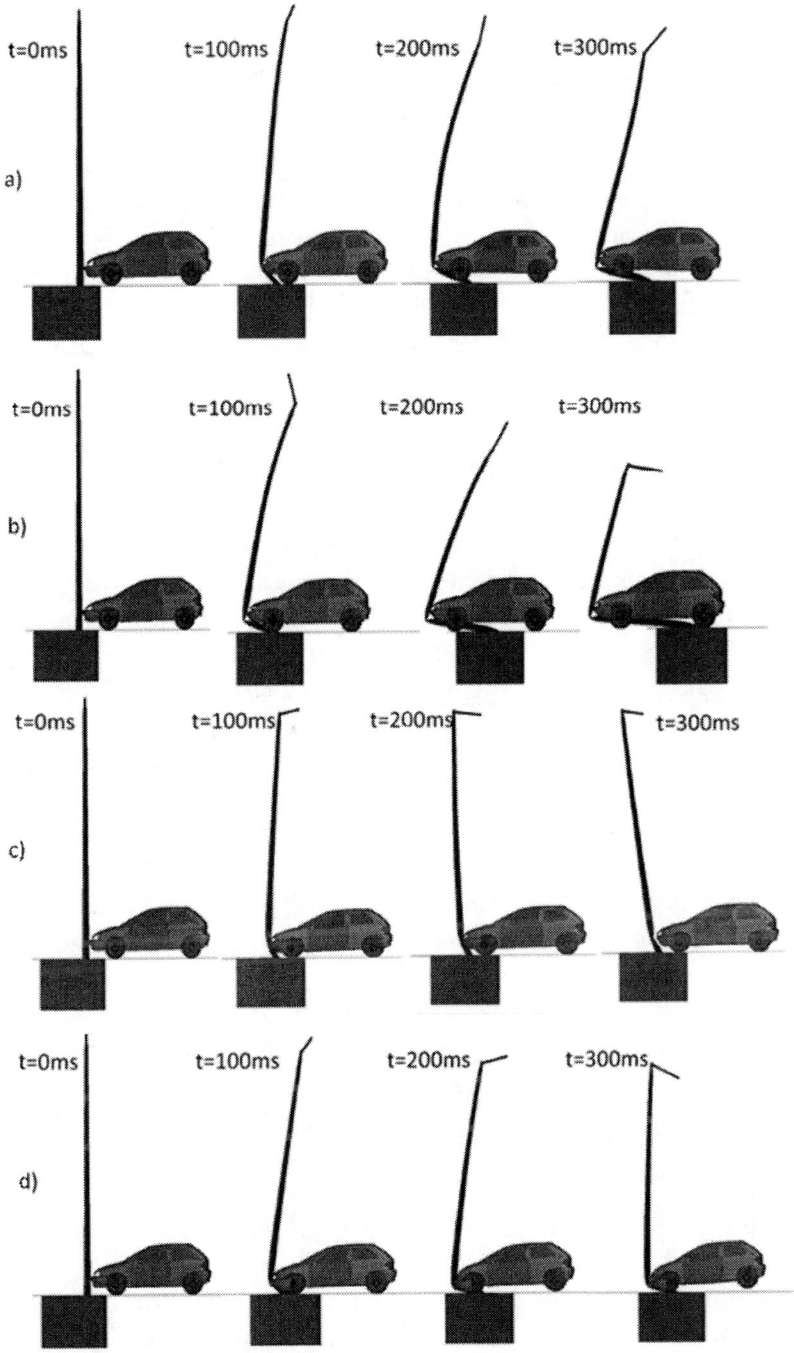

Fig. 3. Qualitative analysis of a collision with lighting column: a) Composite material – velocity 35km/h, b) Composite material – velocity 50km/h, c) Aluminum material – velocity 35km/h, d) Aluminum material – velocity 35km/h.

Based on the results shown on Fig. 3 can be observed that construction made of aluminum absorb more kinetic energy during the collision than construction made of composite material.

4 Summary

Based on the result obtained from numerical simulation passive safety coefficients ASI and THIV was determined. For lighting column made of aluminum was observed that value of ASI coefficient is almost three times higher than for the lighting column made of composite material. For second passive safety coefficient THIV for lighting column made of aluminum this coefficient is lower than for construction made of composite material. Presented in this work Qualitative analysis shown that construction made of aluminum absorb more kinetic energy during collision than lighting column made of composite material. Future work will be done on this model to optimize the shape of the lighting column in order to reduce the values of the passive safety coefficients.

References

1. W. Danek, M. Pawlak, Springer Proceedings in Mathematics & Statistics: Dynamical Systems in Applications, **249**, 73-84 (2018)
2. W. Danek, D. Gąsiorek, Modelowanie Inżynierskie, **66**, 25-30 (2018)
3. M. Pawlak, Mechanics Research Communications, **77**, 21-28 (2016)
4. Y. Abdel-Nasser, J. Alrajhi, M. Alardhi, K. Alkhulaifi, International Journal of Traffic and Transportation Engineering, **2(5)**, 101-105 (2013)
5. G. Janszen, Safety and Security Engineering, **II**, 347-355 (2007)
6. European Committee for Standardization, EN 12767 Passive safety of support structures for road equipment –Requirements and test methods. Draft proposal of revised EN 12767. 2005-06-02. 30p.28
7. European Committee for Standardization, SFS-EN 1317-1 Road Restraint Systems, Part 1: Terminology and General Criteria for Test Methods, Edita Oyj,Helsinki, 1998
8. European Committee for Standardization, SFS-EN 1317-2 Road Restraint Systems, Part 2: Performance Classes, Impact Test Acceptance Criteria and Test Methods for Safety Barriers, Edita Oyj, Helsinki, 1998
9. European Committee for Standardization, SFS-EN 1317-3 Road Restraint Systems, Part 3: Performance Classes, Impact Test Acceptance Criteria and Test Methods for Crash Cushions, Edita Oyj, Helsinki, 2001
10. European Committee for Standardization, SFS-EN 40-3-1 Lighting Columns, Part 3-1: Design and Verification–Specification for Characteristic Loads, Edita Oyj, Helsinki, 2000
11. European Committee for Standardization, SS-EN 40-3-2 Design and verification - Verification by testing, Svensk Standard, 2013
12. European Committee for Standardization, SFS-EN- 40-3-3 Lighting Columns–Part 3-3: Design and Verification–Verification by Calculation, Finnish Standards Association, Helsinki, 2004
13. E. Fasanella, K. Jackson, NASA Langley Research Center (2009)

The Gao beam under a moving inertial load and harmonic compression

Bartłomiej Dyniewicz[1,*] and *Czesław* Bajer[1,**]

[1]Institute of Fundamental Technological Research, Polish Academy of Sciences

Abstract. In the present work the dynamics of the system of a mass moving on the beam is investigated in detail numerically in the case of vibrations about a buckled state. The differential equation that describes the motion is strongly nonlinear. Simulations are based on the space-time finite element method. It enabled us easily determine the influence of the moving inertial particle. At the computational stage it becomes a real problem when the mass particle traverses joints of neighbouring elements. The results of representative and interesting computer simulations are enclosed.

1 Introduction

Structural elements that work in nonlinear range behave in a way far from the known for linear ranges. Well known jumps in trusses under finite displacements, natural frequencies varying under axial loads, buckling etc. are good examples. Modern systems can filter entering excitation and respond with selected and prescribed signals. Special structures exhibit non-classical mechanical or rheological properties.These properties can be predetermined to enable developing new metamaterials. Finally special structures that resist to increased loads, reduce the level of vibration or exhibit increased durability. Nonlinear vibration of a compressed beam is here a good example.

The papers devoted to the Gao beam consider the formulation [1–3], analytical and numerical solutions [4–6], contact problems with the Gao beam [7–9] etc. The mathematical analysis of the problem with contact with the proof of the existence of the solution can be found in [10]. Some particular dynamical problems with contact were discussed in [4, 6, 8, 9], where the model, existence of weak solutions, and computer simulations can be found. An interesting problem of the growth of a crack in the Gao beam was studied in [7].

The motion of a point load travelling on a bending element is a typical problem in transportation. Such systems were extensively investigated, however mostly in a classical linear range. In our previous research the noninear Gao beam subjected to a moving massless or inertial load was investigated. The contribution of a nonlinear term together with a traction force exhibited properties unexpected in a linear range.

This paper continues the study of the vibrations of a nonlinear Gao beam that is subjected to a moving mass [11, 12] by adding a oscillating axial force. Such

*Corresponding author: bdynie@ippt.pan.pl
**Corresponding author: cbajer@ippt.pan.pl

problems arise naturally in transportation systems such as trains or trams. This makes the model, which is a coupled system of the Gao nonlinear beam equation and the motion of the mass, considerably more complex.

In the present work the dynamics of the system of a mass moving on the beam was investigated in detail numerically in the case of vibrations about a buckled state. The simulations are based on the space-time finite element method. It enabled us easily determine the influence of the moving inertial particle. At the computational stage it becomes a real problem when the mass particle traverses joints of neighbouring elements. The simple time integration method separated from the spatial discretization results in complexity in matrix formulation of a resulting systems of algebraic equations (see for example [13]). Moreover, both nonstationary discretization of the doman and nonstationary location of parameters describing moving inertia or granulated stiffness can be applied with this approach relatively easily [14, 15]. The results of representative and interesting computer simulations are enclosed.

2 Formulation

Let us consider the pair of differential equations that govern the motion of the Gao beam (Figure 1)

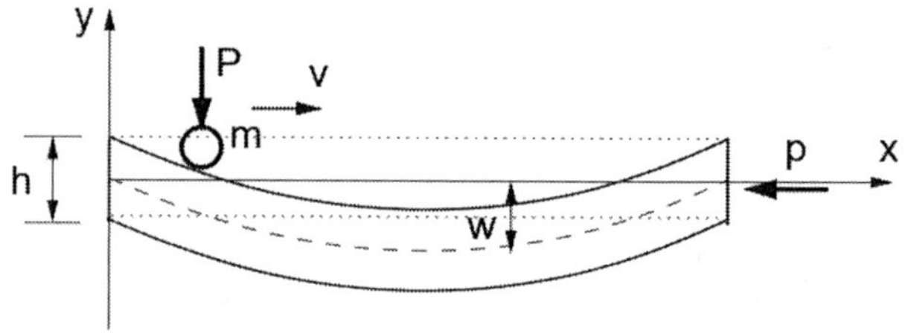

Figure 1. The scheme od the Gao beam.

$$\begin{aligned}&\rho A w_{tt} + EI w_{xxxx} - EA(1+\nu)[((2-\nu)w_x^2 w_{xx} + w_{xx}u_x] = (1-\nu^2)q(x,t)\,,\\&\rho A u_{tt} - EA u_{xx} - EA(1+\nu)w_x w_{xx} = 0\,,\\&0 \le x \le L\,,\quad 0 \le t.\end{aligned} \tag{1}$$

with boundary conditions

$$\begin{aligned}&w(0,t) = w(L,t) = 0, \qquad w_{xx}(0,t) = w_{xx}(L,t) = 0\,,\\&u(0,t) = 0\,,\quad u(L,t) = 0.\end{aligned} \tag{2}$$

and initial conditions

$$w(x,0) = 0\,,\quad w_t(x,0) = 0 \;\text{ and }\; u(x,0) = 0\,,\quad u_t(x,0) = 0\,. \tag{3}$$

The first equation in (1) describes the bending of the beam, similar to the Bernoulli-Euler type, with a non-linear term proportional to the axial stiffness and in general to the deflection. The second equation relates to the axial motion. $q(x,t)$ is the distributed external transverse load.

The classical form of the Gao nonlinear beam equation is extended with the coupling of the beam motion with the moving transverse inertial load. The setting is as follows. $w(x,t)$ and $u(x,t)$ are transverse and longitudinal displacements, respectively, ρ is the mass density, A – cross section area, I – inertia moment of the cross section, E – Young modulus, and ν – Poissone coefficient. The lengths of the beam is L.

A derivation of the Gao beam model can be found in [1, 16], and related models can be found in the references in the introduction. The beam is clamped at both ends ($x = 0, L$) and a horizontal traction p that may depend on time, acts at the end $x = L$. When the traction is tensile ($p < 0$) the beam behaves essentially as the Euler-Bernoulli beam. However, when the traction is compressive and sufficiently large, say $p > p^* > 0$, for some threshold p^*, the beam exhibits two buckled states that are stable and the zero steady state that is unstable. This allows for vibrations about each one of the two buckled states. For the sake of generality, in the formulation we allow the traction $p = p(t)$ to be time dependent and possibly periodic.

We can reduce the system (1) to a single equation. Moreover, the axial vibration can be replaced with constant in length compression, however varying in time. The term $\rho A u_{tt}$ was then put to zero. The given moving mass m, the traction p, the point load P, the mass velocity v and the initial data w_0 and v_0, allows to write the alternative single equation with the moving mass term in the following form

$$\rho w_{tt} + k w_{xxxx} - (a w_x^2 - \overline{\nu} p) w_{xx} = \delta(x - \mathsf{v}t)P - \delta(x - \mathsf{v}t) m w_{tt}(\mathsf{v}t, t), \tag{4}$$

$$w(0,t) = w_x(0,t) = 0, \tag{5}$$

$$w(L,t) = w_x(L,t) = 0, \tag{6}$$

$$w(x,0) = w_0(x), \quad w_t(x,0) = v_0(x). \tag{7}$$

Here ρ is the mass density per unit length, $k = 2h^3 E_Y / 3(1-\nu^2)$, $\overline{\nu} = (1+\nu)$, and $a = 3hE$, h is the heigh of the cross-section, ν and E are the Poisson ratio and the Young modulus, respectively. P is the vertical point force, m is the inertia of the material point accompanying the force, and v is the velocity of the travelling inertial load (P, m).

We note here that the vertical acceleration of the mass particle is given by the Renaudot formula [17],

$$w_{tt}(\mathsf{v}t, t) = \left[w_{tt} + 2(\mathsf{v} + \dot{\mathsf{v}}t) w_{xt} + (\mathsf{v} + \dot{\mathsf{v}}t)^2 w_{xx} + (\ddot{\mathsf{v}}t + 2\dot{\mathsf{v}}) w_x \right]\Big|_{x=\mathsf{v}t} . \tag{8}$$

where the derivatives on the right-hand side are evaluated at $x = \mathsf{v}t$, and here and below a dot above a function indicates its time derivative. For the sake of simplicity, we assume that the velocity v is constant, hence, the expression reduces to

$$w_{tt}(\mathsf{v}t, t) = \left(w_{tt} + 2\mathsf{v} w_{xt} + \mathsf{v}^2 w_{xx} \right)\Big|_{x=\mathsf{v}t} . \tag{9}$$

The system is nonlinear and so it is natural to consider a weak or variational formulation.

3 Numerical model

The differential equation (4) allows us to write the virtual work in the finite element as follows

$$\Pi = \int_\Omega v^* \left\{ \rho w_{tt} + k w_{xxxx} - a w_x^2 w_{xx} + \overline{\nu} p w_{xx} + \delta(x - \mathsf{v}t) m w_{tt}(\mathsf{v}t, t) - \delta(x - \mathsf{v}t) P \right\} dx dt, \tag{10}$$

where the virtual velocity v^* with the parameter α is given by

$$v^*(x,t) = \delta(t-\alpha h)\,(\mathbf{N}\mathbf{v}^*)^T\,. \tag{11}$$

We use this form, chosen from the various possible functions of virtual velocity distributions, because of the simplicity of the time integration of the energy functional Π and simple stability control. Here, $\alpha \in [0,1]$ is a parameter that defines the equilibrium point in the time layer (similarly to β in the Newmark time integration scheme). It affects the accuracy and stability of the resulting time integrating scheme (for details see [18]). Integrating, formally, by parts the virtual work expression (10) leads to the following expression

$$\begin{aligned}\Pi =& k\int_\Omega v^*_{xx} w_{xx}\,\mathrm{d}\Omega + \int_\Omega v^*\,\{\rho \mathsf{v}_t - a w_x^2 w_{xx} + \overline{\nu} p w_{xx} \\ &+ \delta(x-vt)m(v_t + 2\mathsf{v}v_x + \mathsf{v}^2 w_{xx})\big|_{x=vt} - \delta(x-vt)P\}\,\mathrm{d}x\,\mathrm{d}t.\end{aligned} \tag{12}$$

To proceed with the numerical scheme, we need the following matrices: the element mass and stiffness characteristic matrices are given by

$$\mathbf{M} = \rho\begin{bmatrix} \frac{13b}{35} & \frac{11b^2}{210} & \frac{9b}{70} & \frac{-13b}{420} \\ \frac{11b^2}{210} & \frac{b^3}{105} & \frac{13b^2}{420} & \frac{-b^3}{140} \\ \frac{9b}{70} & \frac{13b^2}{420} & \frac{13b}{35} & \frac{-11b^2}{210} \\ \frac{-13b^2}{420} & \frac{-b^3}{140} & \frac{-11b^2}{210} & \frac{b^3}{105} \end{bmatrix}, \tag{13}$$

and

$$\mathbf{K} = k\begin{bmatrix} \frac{12}{b^3} & \frac{6}{b^2} & \frac{-12}{b^3} & \frac{6}{b^2} \\ \frac{6}{b^2} & \frac{4}{b} & \frac{-6}{b^2} & \frac{2}{b} \\ \frac{-12}{b^3} & \frac{-6}{b^2} & \frac{12}{b^3} & \frac{-6}{b^2} \\ \frac{6}{b^2} & \frac{2}{b} & \frac{-6}{b^2} & \frac{4}{b} \end{bmatrix} + \overline{\nu}p\begin{bmatrix} \frac{6}{5b} & \frac{11}{10} & \frac{-6}{5b} & \frac{1}{10} \\ \frac{1}{10} & \frac{2b}{15} & \frac{-1}{10} & \frac{-b}{30} \\ \frac{-6}{5b} & \frac{-1}{10} & \frac{6}{5b} & \frac{-11}{10} \\ \frac{1}{10} & \frac{-b}{30} & \frac{-1}{10} & \frac{2b}{15} \end{bmatrix}. \tag{14}$$

To deal with the nonlinearity, we use a term in which the nonlinearity is frozen at the time t so that $\mathbf{w}^t$ and $\mathbf{v}^t$ are known,

$$\mathbf{K}_n = a\int_0^b \mathbf{N}^T\left[\mathbf{N}_x\mathbf{w}^t + (\alpha - 0.5\alpha^2)h\mathbf{N}_x\mathbf{v}^t + 0.5\alpha^2 h\mathbf{N}_x\mathbf{v}^{t+h}\right]^2 \mathbf{N}_{xx}\,\mathrm{d}x. \tag{15}$$

The explicit matrix form of $\mathbf{K}_n$ is provided in the Appendix.

The general matrices representing the moving mass were derived by using linear shape functions as described in [19],

$$\mathbf{M}_m = m\begin{bmatrix} (1-\kappa)^2 & 0 & \kappa(1-\kappa) & 0 \\ 0 & 0 & 0 & 0 \\ \kappa(1-\kappa) & 0 & \kappa^2 & 0 \\ 0 & 0 & 0 & 0 \end{bmatrix}, \quad \mathbf{C}_m = \frac{2m\mathsf{v}}{b}\begin{bmatrix} \kappa-1 & 0 & 1-\kappa & 0 \\ 0 & 0 & 0 & 0 \\ -\kappa & 0 & \kappa & 0 \\ 0 & 0 & 0 & 0 \end{bmatrix},$$

$$\mathbf{e}^t_m = \frac{m\mathsf{v}}{bh}\begin{bmatrix} (1-\kappa)(\mathsf{w}_2^{i+1} - \mathsf{w}_2^i - \mathsf{w}_1^{i+1} + \mathsf{w}_1^i) \\ 0 \\ \kappa(\mathsf{w}_2^{i+1} - \mathsf{w}_2^i - \mathsf{w}_1^{i+1} + \mathsf{w}_1^i) \\ 0 \end{bmatrix}. \tag{16}$$

The upper indices i and $i+1$ are the numbers of neighboring spatial elements traversed by the inertial particle. The lower indices 1 and 2 correspond to left and right nodes in each neighbouring element. The differences are equal to zero except the case the mass travels from one element to another during the time interval Δt. The moving force vector is given as

$$\mathbf{F}^t = P \begin{bmatrix} 1-\kappa \\ 0 \\ \kappa \\ 0 \end{bmatrix}, \tag{17}$$

where $\kappa \in [0,1]$ is parameter that describes current position of the moving load in finite element. Finally, the minimization of the functional in (12) and the above matrices lead to our numerical scheme, which is of the marching in time type.

Computations were performed according to the algorithm: given $\mathbf{w}^t$ and $\mathbf{v}^t$, as well as $\mathbf{e}_m^t$ and $\mathbf{F}^t$, find $\mathbf{w}^{t+h}$ and $\mathbf{v}^{t+h}$ by solving the system

$$\begin{aligned} &\left[\mathbf{M}/h + \mathbf{M}_m/h + \alpha\mathbf{C}_m + 0.5\alpha^2 h(\mathbf{K}-\mathbf{K}_n)\right]\mathbf{v}^{t+h} \\ &-\left[\mathbf{M}/h + \mathbf{M}_m/h - (1-\alpha)\mathbf{C}_m \right. \\ &\left. -(\alpha - 0.5\alpha^2)h(\mathbf{K}-\mathbf{K}_n)\right]\mathbf{v}^t \\ &+(\mathbf{K}-\mathbf{K}_n)\mathbf{w}^t + \mathbf{e}_m^t - \mathbf{F}^t = 0\,, \end{aligned} \tag{18}$$

and

$$\mathbf{w}^{t+h} = \mathbf{w}^t + h[(1-\alpha)\mathbf{v}^t + \alpha\mathbf{v}^{t+h}]\,. \tag{19}$$

The data set for a steel beam used in the simulations was as follows:
L=100 [cm], β=0.015, E=2.07 [g/cm/μs^2], ν=0.3, $\rho = 7.7\frac{\beta^2 L^2}{\pi}$ =5.5175 [g/cm], $A=\beta^2L^2/\pi$, $I=\beta^4 l^4/4/\pi^3$, $k = 0.57\frac{\beta^4 L^4}{4\pi^3}$ =0.0233 [g cm^3/μs^2], $\mathsf{v} = \frac{\pi\sqrt{k/\rho}}{L}\alpha_v$ = 0.002 α_v [cm/μs], m = 100ρL =55175 [g], $P = mg$ = $5.4071 * 10^{-5}$ [g cm/μs^2], p^* =0.71 [g cm/μs^2], g=$9.8\cdot 10^{-10}$ [cm/μs^2].

4 Results

Numerous interesting results can be depicted since the response of the system depends on several parameters, mainly the stiffness and inertia, beam compression p, the load velocity, its inertia, and finally the resulting magnitude of deflection. Figure 2 depicts displacements under the inertial load during the passage of a series of loads. The distance between successive inertial loads was equal to the beam length. Four passaged were depicted, each line with increasing compression p, constant in time. In the case of zero compression the first stage exhibits significant deflection under the load while in successive stages the inertia of the load stabilizes the trajectory and pulls the beam toward zero deflection. The increased p makes the beam soft and sensitive to the initial stage of vibration. In the extreme cases the beam deflects to the opposite side to the load direction. It is visible in Figure 3. Significant differences between dynamical responses of the beam under constant and oscillating axial compression is depicted in Figure 4. In the case of a moving load the oscillation of p reduces deflection under the load while the influence of a moving inertia increases amplitudes. Differences in both diagrams in Figure 4 are significant and the phenomenon should be investigated further.

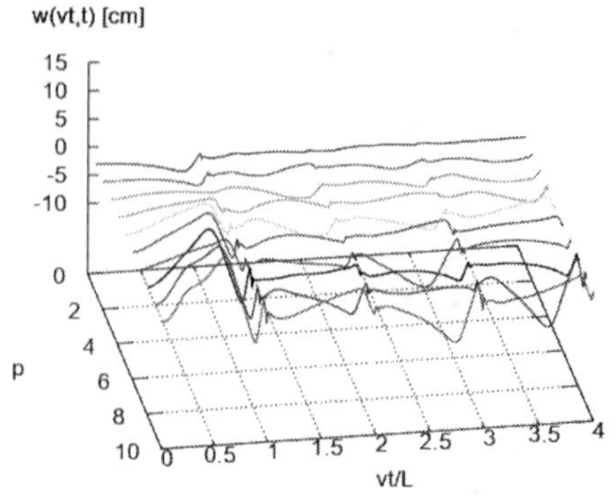

Figure 2. Trajectory of the contact point with the Gao beam subjected to a sequence of inertial forces for increasing p.

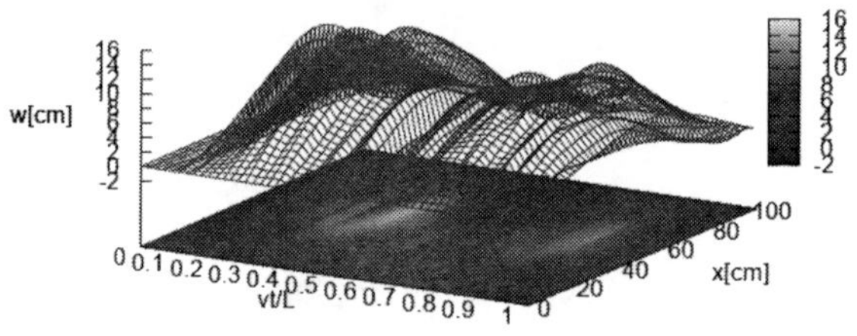

Figure 3. Displacements in time for α_v=0.05, β=0.15 for inertial load with p=4.5.

5 Conclusions

The compressed beam that works in a strongly nonlinear range and subjected to a moving load exhibits new properties. The deflection depends on several parameters, mainly on the compressive force p and the contribution of the inertial part of the load. The most important feature is the possibility of deflection in opposite direction to the transverse load. It can occur due to the rotational inertia of the part of the beam just subjected to the inertial point load. The strongly compressed beam then jumps out and is held in the upper position.

Acknowledgment This research has been supported within the projects UMO-2015/17/B/ST8/03244 and UMO-2017/26/E/ST8/00532 funded by the Polish National Science Centre, which is gratefully acknowledged by the authors.

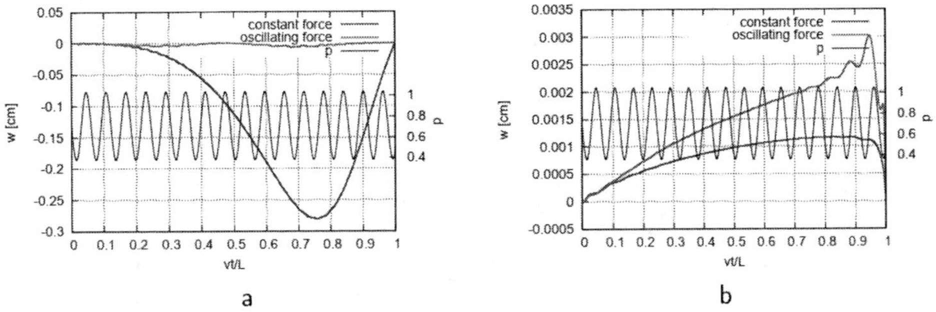

Figure 4. Displacements in time of the contact point under the travelling force (a) and the mass (b).

References

[1] D.Y. Gao. Nonlinear elastic beam theory with application in contact problems and variational approaches. *Mechanics Research Communications*, 23(1):11–17, 1996

[2] D.Y. Gao and D.L. Russell. An extended beam theory for smart materials applications: II. Static formation problems. *Appl. Math. Optim.*, 38(1):69–94, 1998

[3] D.L. Russell and L.W. White. A nonlinear elastic beam system with inelastic contact constraints. *Appl. Math. Optim.*, 46:291–312, 2002

[4] K.T. Andrews, M.F. M'Bengue, and M. Shillor. Vibrations of a nonlinear dynamic beam between two stops. *Discrete and Continuous Dynamical System (DCDS-B)*, 12(1):23–38, 2009

[5] M.F. M'Bengue and M. Shillor. Regularity result for the problem of vibrations of a nonlinear beam. *Electron. J. Diff. Eqns.*, 27:1–12, 2008

[6] K.T. Andrews, Y. Dumont, M.F. M'Bemgue, J. Purcell, and M. Shillor. Analysis and simulations of a nonlinear dynamic beam. *Appl. Math. Physics (ZAMP)*, 63(6):1005–1019, 2012

[7] K.L. Kuttler, J. Purcell, and M. Shillor. Analysis and simulations of a contact problem for a nonlinear dynamic beam with a crack. *Quarterly J. Mech. Appl. Math.*, 2011 DOI: 10.1093/qjmam/hbr018.

[8] J. Ahn, K.L. Kuttler, and M. Shillor. Dynamic contact of two Gao beams. *Electron. J. Diff. Equ.*, 194:1–42, 2012

[9] K.T. Andrews, K.L. Kuttler, and M. Shillor. *Dynamic Gao beam in contact with a reactive or rigid foundation*, volume 33, pages 225–248. 2015

[10] M.F. M'Bengue. *Analysis of a Nonlinear Dynamic Beam with Material Damage or Contact.* PhD thesis, Oakland University, 2008

[11] C.I. Bajer, B. Dyniewicz, and M. Shillor. A Gao beam subjected to a moving inertial point load. *Mathematics and Mechanics of Solids*, 23(3):461–472, 2018

[12] B. Dyniewicz, C.I. Bajer, K.L. Kuttler, and M. Shillor. Vibrations of a Gao beam subjected to a moving mass (in print). *Nonlinear Analysis: Real World Applications*, 2019

[13] C.I. Bajer and B. Dyniewicz. *Numerical analysis of vibrations of structures under moving inertial load.* Springer, 2012

[14] C. I. Bajer. Adaptive mesh in dynamic problem by the space–time approach. *Computers and Structures*, 33(2):319–325, 1989

[15] B. Dyniewicz, D. Pisarski, and C. Bajer. Vibrations of a mindlin plate subjected to a pair of inertial loads moving in opposite directions. *Journal of Sound and Vibration*, 386:265–282, 2017

[16] D.Y. Gao. Finite deformation beam models and triality theory in dynamical post-buckling analysis. *Intl. J. Non-Linear Mechanics*, 35:103–131, 2000

[17] A. Renaudot. Etude de l'influence des charges en mouvement sur la resistance, des ponts metallique a poutres droites. *Annales des Ponts et Chausses*, 1:145–204, 1861

[18] C. I. Bajer and B. Dyniewicz. Virtual functions of the space-time finite element method in moving mass problems. *Computers and Structures*, 87:444–455, 2009

[19] B. Dyniewicz. Space-time finite element approach to general description of a moving inertial load. *Finite Elem. Anal. and Des.*, 62:8–17, 2012

A Appendix

$$
\mathbf{K}_n = \begin{bmatrix}
-2(b^2(8q^2-3qs+s^2)+3b(p-r)(3s-4q)+12(p^2-2pr+r^2))/(35b^3) & -(b^2(139q^2-50qs+25s^2)+12b(p-r)(17s-13q)+396(p^2-2pr+r^2))/(420b^2) & 2(b^2(8q^2-3qs+s^2)+3b(p-r)(3s-4q)+12(p^2-2pr+r^2))/(35b^3) & (b^2(53q^2-22qs-s^2)+12b(p-r)(s-11q)-108(p^2-2pr+r^2))/(420b^2) \\
-(b^2(3q^2-2qs+s^2)+12b(p-r)(s-q)+12(p^2-2pr+r^2))/(140b^2) & (b^2(2q^2-qs+s^2)+3b(p-r)(3s-q)+18(p^2-2pr+r^2))/(105b) & (b^2(3q^2-2qs+s^2)+12b(p-r)(s-q)+12(p^2-2pr+r^2))/(140b^2) & -(b^2(q^2-2qs-s^2)+24bq(r-p)-36(p^2-2pr+r^2))/(420b) \\
2(b^2(q^2-3qs+8s^2)+3b(p-r)(3q-4s)+12(p^2-2pr+r^2))/(35b^3) & -(b^2(q^2+22qs-53s^2)+12b(p-r)(11s-q)+108(p^2-2pr+r^2))/(420b^2) & -2(b^2(q^2-3qs+8s^2)+3b(p-r)(3q-4s)+12(p^2-2pr+r^2))/(35b^3) & (b^2(25q^2-50qs+139s^2)+12b(p-r)(17q-13s)+396(p^2-2pr+r^2))/(420b^2) \\
-(b^2(q^2-2qs+3s^2)+12b(p-r)(q-s)+12(p^2-2pr+r^2))/(140b^2) & (b^2(q^2+2qs-s^2)+24bs(p-r)+36(p^2-2pr+r^2))/(420b) & (b^2(q^2-2qs+3s^2)+12b(p-r)(q-s)+12(p^2-2pr+r^2))/(140b^2) & -(b^2(q^2-qs+2s^2)+3b(p-r)(3q-s)+18(p^2-2pr+r^2))/(105b)
\end{bmatrix},
$$

where

$$
\begin{aligned}
p &= \mathrm{w}_1 + (\alpha - 0.5\alpha^2)h\mathrm{v}_1 + 0.5\alpha^2 h\mathrm{v}_3\,,\\
q &= \psi_1 + (\alpha - 0.5\alpha^2)h\dot{\psi}_1 + 0.5\alpha^2 h\dot{\psi}_3\,,\\
r &= \mathrm{w}_2 + (\alpha - 0.5\alpha^2)h\mathrm{v}_2 + 0.5\alpha^2 h\mathrm{v}_4\,,\\
s &= \psi_2 + (\alpha - 0.5\alpha^2)h\dot{\psi}_2 + 0.5\alpha^2 h\dot{\psi}_4\,.
\end{aligned}
$$

Dynamic stress intensity factors of interacting branched cracks

Piotr Fedeliński[1,*]

[1]Silesian University of Technology, 44-100 Gliwice, Poland

Abstract. In the present work the boundary element method (BEM) is applied to analysis of statically and dynamically loaded infinite plates with multiple stationary branched cracks. The material of the plates is linear-elastic, homogenous and isotropic. In the applied BEM approach the displacement and traction boundary integral equations are used simultaneously for nodes on crack surfaces. Contrary to the finite element method (FEM) in the BEM numerical solutions are obtained by discretization of external boundaries and crack surfaces. The dynamic problem is solved by using the Laplace transform method and the solution in the time domain is computed by the Durbin numerical inversion method. Numerical examples of multiple branched cracks in infinite plates subjected to static and dynamic loadings are presented. An influence of orientation, distances between cracks and the number of cracks on static and dynamic stress intensity factors (SIF) is studied.

1 Introduction

Branched cracks may be material defects that arise during the manufacture of materials or structures and as a result of exploitation. Single cracks can branch out during propagation, especially under dynamic loading. If there are many cracks in the material then they interact with each other causing a change in the stress fields. This interaction depends on the orientation, distance and number of cracks. The purpose of the work is to determine the influence of crack interaction on static and dynamic stress intensity factors (SIF), which characterize stresses near the tips of cracks.

Daux et al. [1] presented the extended finite element method (X-FEM) to analysis of cracks with multiple branches and cracks emanating from holes. A standard displacement approximation was enriched by incorporating additional discontinuous functions. The method allows the modelling of discontinuities independently of the mesh. Raffie et al. [2] investigated bifurcation and trifurcation of fast running cracks under various biaxial loading conditions. Arbitrary curvilinear crack propagations were analyzed by the time-domain BEM. Branching of cracks was controlled by the opening mode SIF and velocity and direction of crack growth by the maximum circumferencial stress at the crack tip. Numerical solutions were compared with experimental results.

* Corresponding author: piotr.fedelinski@polsl.pl

In the present work the boundary element method (BEM) is applied to analysis of statically and dynamically loaded branched cracks in infinite plates. The crack problem is solved by the dual BEM which was developed for static loadings by Portela et al. [3] and for dynamical loadings by Fedeliński et al. [4]. An overview of different BEM approaches in dynamic fracture mechanics was presented by Fedeliński [5]. The application of the method for single branched cracks was shown by Fedeliński in [6]. In this work the Laplace transform method is applied to analysis of interacting branched cracks in infinite plates.

In the dual BEM only boundaries of the body and crack surfaces are divided into boundary elements. The variations of boundary coordinates, displacements and tractions are interpolated using shape functions and nodal values. The relations between boundary displacements and tractions are expressed by the displacement and traction boundary integral equations. The displacement equation is applied for nodes on the external boundary and both equations for nodes on crack surfaces. The time dependent problems are solved using the Laplace transform method and the solution in the time domain is computed by the Durbin numerical inversion method. In the dual BEM the boundary displacements and tractions are computed directly. Static stress intensity factors (SIF) are calculated using the path independent J-integral and the dynamic SIF using crack opening displacements.

2 Numerical examples – branched cracks in an infinite domain

Static and dynamic stress intensity factors (SIF) of multiple branched cracks in an infinite domain are analyzed. The material of the plate has the following elastic properties: the Young modulus E=200 GPa, the Poisson ratio ν=0.3, the mass density ρ=8000 kg/m^3 and the plate is in the plane strain conditions. The velocity of the longitudinal wave is c_1=5801 m/s and the velocity of the shear wave is c_2=3101 m/s. The plate is loaded by the Heaviside time dependent uniformly distributed tension p applied at the initial time t=0. The static tension has the constant value p. The numerical solution is obtained using 50 Laplace parameters and the time step is Δt=0.5 μs. In order to model an infinite domain a large square plate having dimensions h=w=15 cm is considered, as shown in Fig. 1a. The plate is constrained along the right edge and loaded along the left edge. The branched cracks are situated near the center of the plate O (Fig. 1b). The cracks have three branches of length a=1 cm and the angle between the branches is $2/3\pi$. The distance of crack centers from the point O is l, unless otherwise defined. The crack tips of the horizontal and inclined branches are denoted by A and B, respectively. The length of the cracks in the vertical direction is $2c_2$. The SIF are normalized with respect to the static SIF K_o=$p(\pi c_2)^{1/2}$ for the straight crack of length $2c_2$ in an infinite domain loaded by the remote tension p. The external boundary of the square plate is divided into 200 boundary elements and each branched crack into 60 boundary elements having the same length.

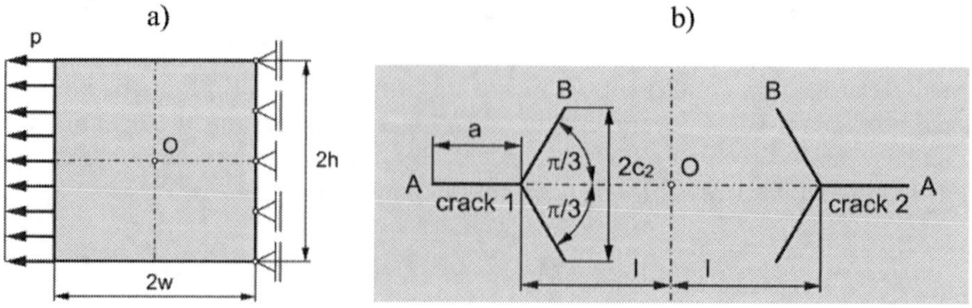

Fig. 1. Branched cracks in a square plate: a) dimensions of the plate and boundary conditions, b) dimensions of branched cracks.

2.1 Different orientations of two branched cracks in an infinite domain

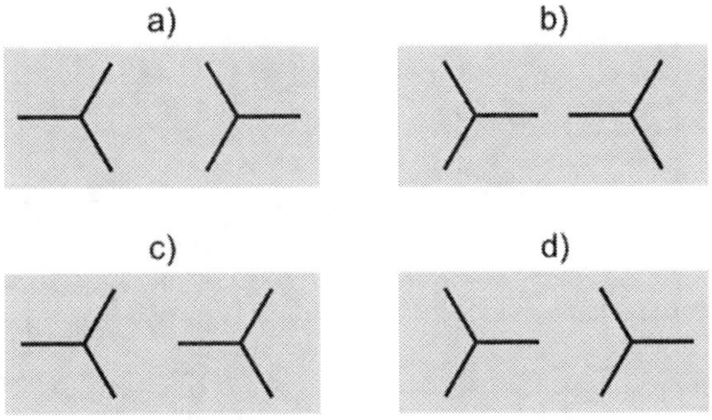

Fig. 2. Orientation of two branched cracks in an infinite domain.

Four different orientations of cracks shown in Fig. 2 are considered. The distance of crack centers from the point O is l=1.5 cm. The normalized static SIF K/K_o are given in Table 1. Because of the symmetry of the cracks the results are not shown for the crack 2 for the case (a) and (b) and for the case (d). The results are compared with the solutions for the single crack. The SIF $K_I/K_o(A)$ are very small and the second crack has small influence on the values. The SIF $K_I/K_o(B)$ has smaller values when the crack tips B are in a small distance (case (a) and (c)). The SIF $K_{II}/K_o(B)$ for the first crack is not sensitive to the orientation of cracks. The SIF $K/K_o(B)$ for the inclined branch B in each case are smaller than for the single crack.

Table 1. Normalized SIF for different orientations of two branched cracks in an infinite domain.

orientation	crack 1			crack 2		
	$K_I(A)$	$K_I(B)$	$K_{II}(B)$	$K_I(A)$	$K_I(B)$	$K_{II}(B)$
a	0.029	0.628	0.490	-	-	-
b	0.051	0.727	0.489	-	-	-
c	0.028	0.674	0.500	0.065	0.717	0.464
single	0.016	0.754	0.541	-	-	-

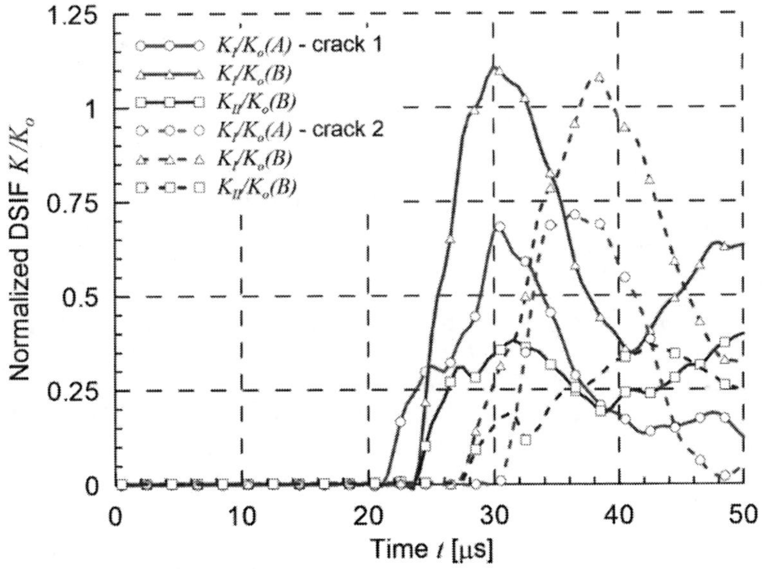

Fig. 3. Normalized DSIF K/K_o for the orientation of cracks (a).

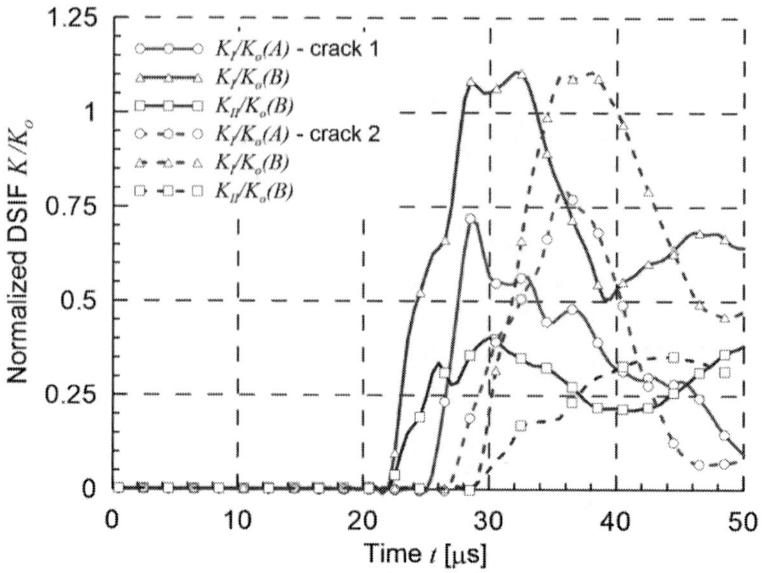

Fig. 4. Normalized DSIF K/K_o for the orientation of cracks (*b*).

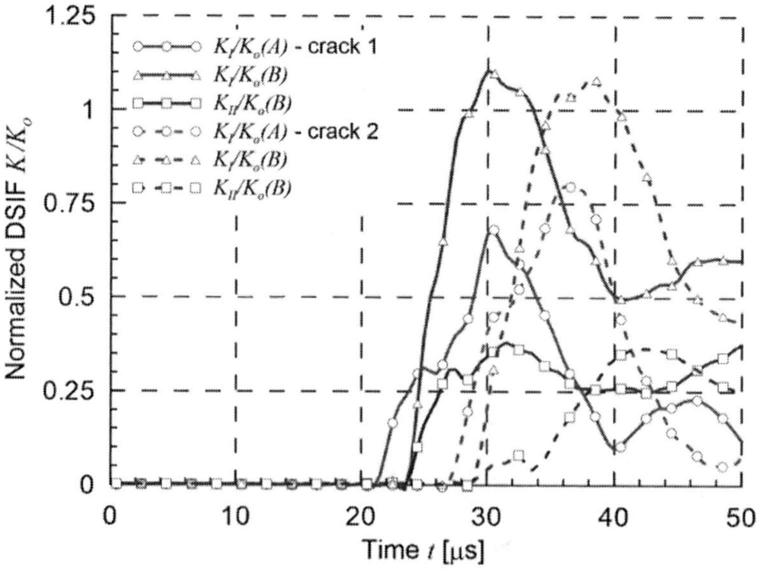

Fig. 5. Normalized DSIF K/K_o for the orientation of cracks (*c*).

The normalized dynamic DSIF K/K_o for different orientations are presented in Fig. 3-6. The longitudinal wave propagates from the loaded boundary to the crack tip A in cases (a) and (c) 21.5 μs and to the crack tip B in cases (b) and (d) 22.4 μs. During this time the cracks are at rest and the DSIF are zero. The time dependences of DSIF for each configuration of cracks are different but the maximum values are similar. The static SIF for the horizontal branches are small but the dynamic DSIF have significant values. The normalized dynamic DSIF $K_I/K_o(B)$ are larger than $K_I/K_o(A)$.

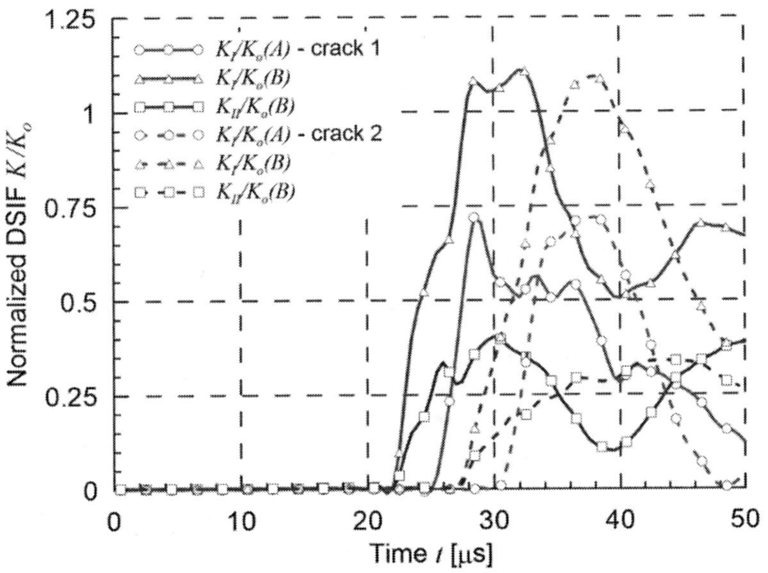

Fig. 6. Normalized DSIF K/K_o for the orientation of cracks (*d*).

2.2 Different distance between two branched cracks in an infinite domain

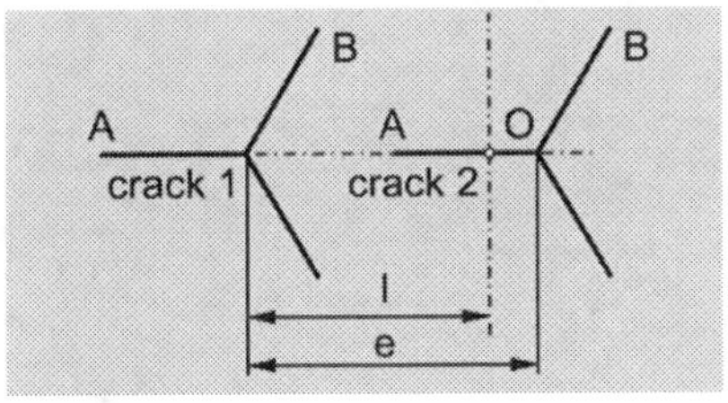

Fig. 7. Different distances between two branched cracks in an infinite domain.

The influence of the distance *e* between crack centers on SIF for the configuration shown in Fig. 7 is studied. The distances are *e*=2.0, 2.5, 3.0, 3.5 cm. The normalized static SIF K/K_o are given in Table 2. The results are compared with the solution for the single crack. When the distance increases the SIF for the inclined branches increase.

Table 2. Normalized SIF for different distances between two branched cracks in an infinite domain.

distance	crack 1			crack 2		
e/l	$K_I(A)$	$K_I(B)$	$K_{II}(B)$	$K_I(A)$	$K_I(B)$	$K_{II}(B)$
2.0/1.5	0.032	0.614	0.488	0.014	0.700	0.416
2.5/1.5	0.031	0.649	0.493	0.068	0.709	0.441
3.0/1.5	0.028	0.674	0.500	0.065	0.717	0.464
3.5/1.5	0.027	0.692	0.506	0.055	0.722	0.480
single	0.016	0.754	0.541	-	-	-

The normalized dynamic DSIF K_I/K_o for the crack tips *A* and *B* are shown in Fig. 8 and 9, respectively. The DSIF for the horizontal branch *A* are larger for the second crack than for the first crack. The distance between the cracks have the significant influence on the $K_I/K_o(A)$ for the first crack for times larger than 36 μs (Fig. 8). The maximal values of DSIF $K_I/K_o(B)$ for the inclined branches are similar (Fig. 9).

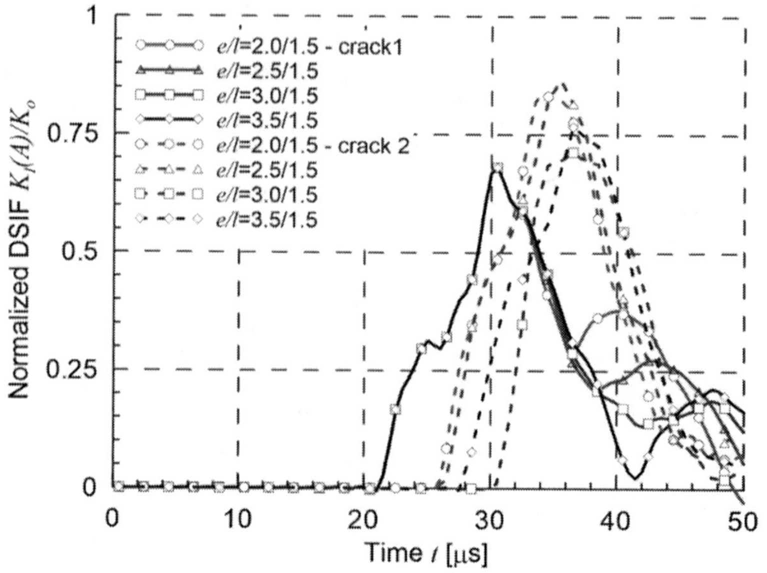

Fig. 8. Normalized DSIF $K_I/K_o(A)$ for the horizontal branches.

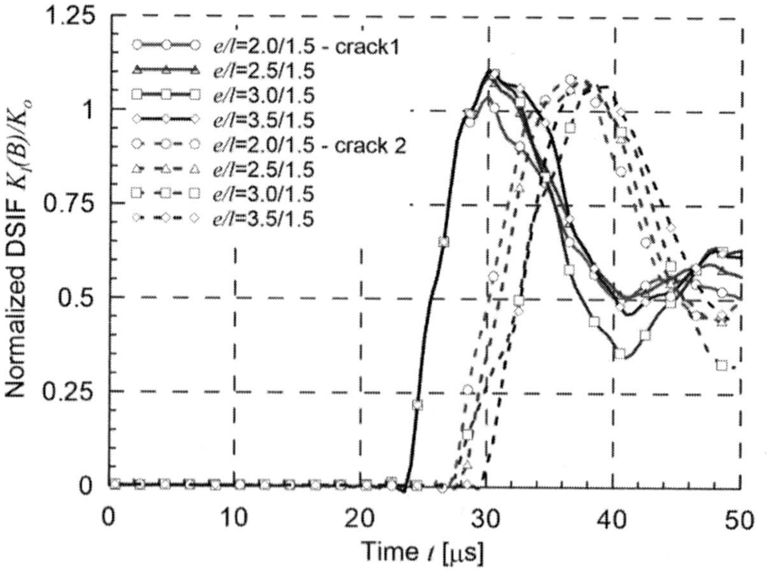

Fig. 9. Normalized DSIF $K_I/K_o(B)$ for the inclined branches.

2.3 Different number of branched cracks in an infinite domain

An influence of different number of interacting cracks shown in Fig. 10 on SIF is studied. The number of cracks varies from one to four. The distance between the cracks is e=2.0 cm. The normalized static SIF are given in Table 3. When the number of cracks increases the $K_I/K_o(A)$ increase and $K/K_o(B)$ decreases. The normalized dynamic DSIF for four cracks are shown in Fig. 11 and for the first crack tip A and B for different number of cracks in Fig. 12. It is interesting to note in Fig. 11 that $K_I/K_o(A)$ is gradually increasing for the following cracks.

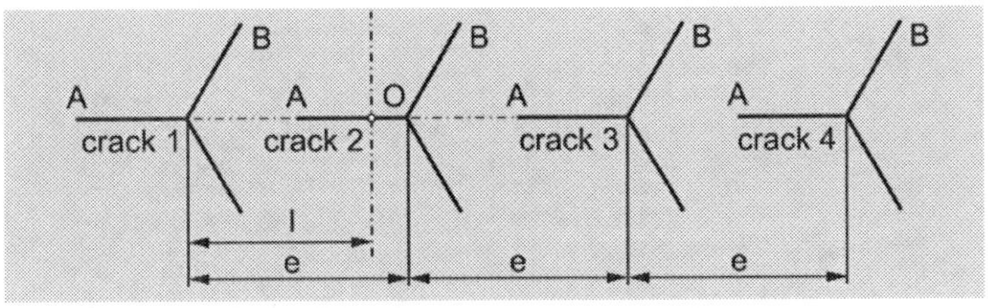

Fig. 10. Different number of branched cracks in an infinite domain.

Table 3. Normalized SIF for different number of branched cracks in an infinite domain.

number of cracks	$K_I(A)$	$K_I(B)$	$K_{II}(B)$
crack 1			
1	0.016	0.754	0.541
2	0.032	0.614	0.488
3	0.041	0.600	0.472
4	0.047	0.595	0.461
crack 2			
2	0.014	0.700	0.416
3	0.036	0.567	0.364
4	0.047	0.554	0.346
crack 3			
3	0.053	0.689	0.392
4	0.076	0.558	0.340
crack 4			
4	0.069	0.684	0.381

3 Conclusions

Multiple branched cracks in an infinite plate subjected to static and dynamic loadings were considered. The influence of orientation, distance and number of cracks on stress intensity factors was studied. Static SIF for branches parallel to the applied loading are small but the dynamic DSIF have significant values. The dynamic opening DSIF for inclined branches with respect to the loading are larger than DSIF for the parallel branches. The dynamic opening DSIF for multiple cracks are the smallest for the parallel branch of the first crack and the largest for the last crack.

Acknowledgement

The scientific research is financed by National Science Centre, Poland, in years 2016-2019, grant no. 2015/19/B/ST8/02629.

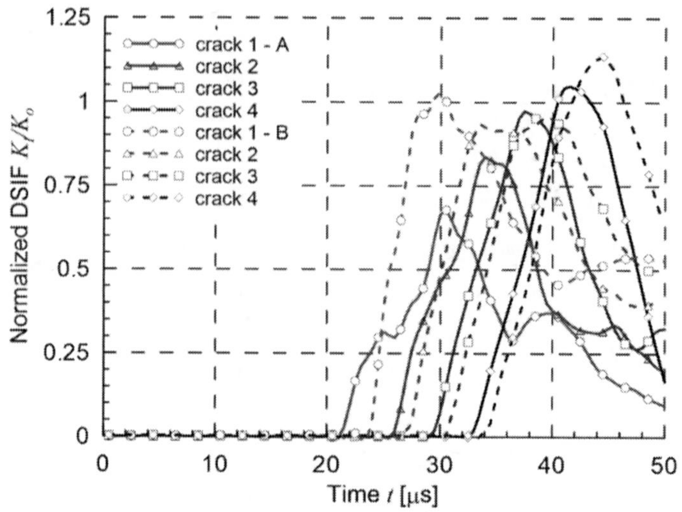

Fig. 11. Normalized DSIF K_I/K_o for four branched cracks in an infinite domain.

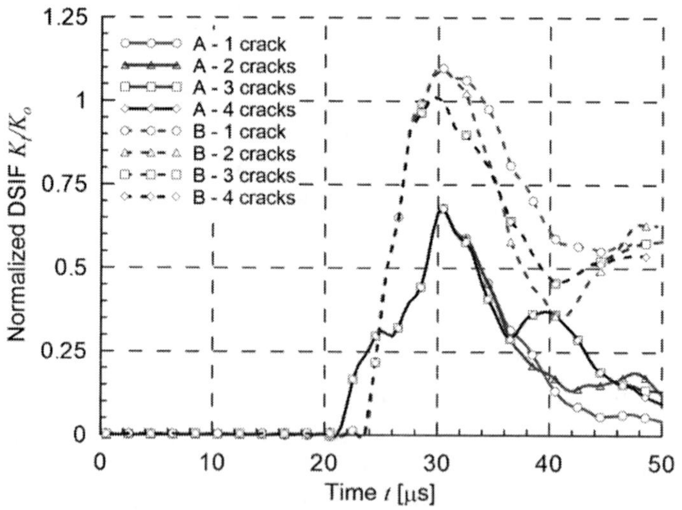

Fig. 12. Normalized DSIF K_I/K_o for the first crack for different number of branched cracks in an infinite domain.

References

1. Ch. Daux, N. Moes, J. Dolbow, N. Sukumar, T. Belytschko, Int. J. Numer. Meth. Engng, **48**, 1741-1760 (2000)
2. S .Raffie, Th. Seelig, D. Gross. Int. J. Frac., **120**, 545-561 (2003).
3. A. Portela, M. H. Aliabadi, D. P. Rooke, Int. J. Numer. Meth. Engng, **33**, 1269-1287 (1992)
4. P. Fedeliński, M. H. Aliabadi, D.P. Rooke, Comp. Struct., **59**, 1021-1031 (1996)
5. P. Fedeliński, Engng Anal. Bound. Elem., **28**, 1135-1147 (2004)
6. P. Fedeliński P., J. Civil Engng, Envir. Arch., **34**, 17-26 (2017)

Application of the two-parameter single shooting method for natural vibration analysis of Euler-Bernoulli beams

Ryszard Hołubowski*

Wroclaw University of Science and Technology, Faculty of Civil Engineering, Wroclaw, Poland

Abstract. The paper presents a new method of calculating the natural frequencies and modes of Euler-Bernoulli beams. The formulated algorithm is based on the fourth order homogeneous ordinary differential equation of motion that allows different support conditions to be considered. The idea of the method is to replace a two-point boundary value problem with an initial value problem. Distinct from the classic approach, where only one unknown parameter is considered, the method enables problems with two parameters to be solved. The proposed two-parameter version of the single shooting method is verified on the example of a non-prismatic simply supported beam and a prismatic propped cantilever beam resting on Winkler foundation and loaded by axial force. The obtained results confirm the correctness and high effectiveness of the presented method.

1 Introduction

Natural vibration analysis is one of the fundamental issues of structural dynamics. Knowledge of eigenfrequencies and eigenmodes allows structures to be protected against resonant states, structural parameters to be identified, and inter alia forced vibrations to be analysed. Therefore, the natural vibration problem is considered by many researchers. In this paper, an originally formulated two-parameter single shooting method has been used to compute the natural frequencies and modes of Euler-Bernoulli beams. The proposed method can be an alternative to the commonly used finite element method due the fact that the beam does not have to be divided into many elements to obtain accurate enough solution, even though the geometry of the beam is very complex.

The standard single shooting method is based on the second order ordinary differential equation and allows a two-point boundary value problem to be reduced to an initial one [1]. In the literature, we can find various modifications of the classical approach. El-Gebeily and Attili [2] proposed an iterative shooting method for a certain class of singular two-point boundary value problems. Pai and Palazotto [3] used a multiple shooting method in the large-deformation analysis of flexible beams. Ha [4] introduced a new shooting technique for non-linear boundary value problems. Ibrahim et al. [5] investigated large amplitude periodic forced vibrations of curved beams under periodic excitation using a modified shooting approach based on the Newmark time marching scheme and arc length

* Corresponding author: ryszard.holubowski@pwr.edu.pl

continuation algorithm. Ribeiro [6] demonstrated an algorithm based on the shooting method to solve finite element equations of motion and to study geometrically non-linear vibrations of beams and plates.

2 Two-parameter single shooting method

The idea of the two-parameter single shooting method will be presented using the example of natural transverse vibrations of a Euler-Bernoulli beam. To obtain natural frequencies and modes, the following partial differential equation that describing the free vibration process is considered [7]

$$\frac{\partial^2}{\partial x^2}\left[EI(x)\frac{\partial^2 w(x,t)}{\partial x^2}\right]+\rho A(x)\frac{\partial^2 w(x,t)}{\partial t^2}+\frac{\partial}{\partial x}\left[p(x)\frac{\partial w(x,t)}{\partial x}\right]+k(x)w(x,t)=0\,, \tag{1}$$

where $EI(x)$ is the bending stiffness, $\rho A(x)$ represents the mass per unit length, $p(x)$ is the axial compressive force, and $k(x)$ describes the stiffness parameter of the Winkler foundation. The solution of Eq. (1) can be written in the form of separated variables

$$w(x,t)=f(x)g(t)\,. \tag{2}$$

Substituting formula (2) into Eq. (1) and introducing a constant ω^2 leads to two ordinary differential equations

$$\ddot{g}(t)+\omega^2 g(t)=0\ , \tag{3.1}$$

$$\begin{aligned}&EI(x)f''''(x)+2EI'(x)f'''(x)+\left[EI''(x)+p(x)\right]f''(x)+\\&+p'(x)f'(x)+\left[k(x)-\omega^2\rho A(x)\right]f(x)=0\,.\end{aligned} \tag{3.2}$$

where $(\,\dot{}\,)=d/dt$ and $(\,)'=d/dx$.
The general solution of Eq. (3.1) has the harmonic form

$$g(t)=C_1\sin\omega t+C_2\cos\omega t\,. \tag{4}$$

Taking into account the following initial conditions

$$g(0)=0 \ \text{ and } \ \dot{g}(0)=\omega\,, \tag{5}$$

we obtain the particular integral

$$g(t)=\sin\omega t \tag{6}$$

and the solution of Eq. (1) in the form

$$w(x,t)=f(x)\sin\omega t\,, \tag{7}$$

in which ω is the circular natural frequency and function $f(x)$ describes the corresponding natural mode. Substituting solution (7) into Eq. (1) gives the ordinary differential equation

$$\frac{d^2}{dx^2}\left[EI(x)\frac{d^2}{dx^2}f(x)\right]-\omega^2\rho A(x)f(x)+\frac{d}{dx}\left[p(x)\frac{d}{dx}f(x)\right]+k(x)f(x)=0\,, \tag{8}$$

which is a starting point for the shooting method. Despite the type of beam static scheme, the conditions imposed on the solution are always determined at different points of the interval of a solution's definition. In most cases, a two-point boundary value problem is considered. Usually, two conditions are specified at the beginning point and two at the end point of the beam. In order to apply one of the numerical integration methods to solve Eq. (8), all initial conditions have to be known. Therefore, the standard single shooting method is insufficient because it allows the replacement of only one condition, which is determined at the end, with the initial condition. In the considered case, apart from two conditions defined at the end, there is an additional parameter ω, which is the sought unknown. The method proposed in this paper enables the raised problem to be solved.

The procedure of the method, explained on the example of beam natural transverse vibration, is as follows:

a) Formulate a fourth-order ordinary differential equation (8) with a natural frequency ω as the parameter and a solution $f(x)$ defined on an interval [a, b].
b) Evaluate values of function $\eta_1(\omega,\alpha) = d^{k_1} f(b) / dx^{k_1}$ to create a first hit surface. The parameter α represents a value of one of two unknown initial conditions, whereas the value of the second one is freely determined due to the fact that every single mode can be arbitrarily normalizing. The order k_1 of the derivative depends on the boundary conditions.
c) Similar to the previous point, build a second hit surface $\eta_2(\omega,\alpha) = d^{k_2} f(b) / dx^{k_2}$ that is based on a second boundary condition.
d) Solve the system of two nonlinear algebraic equations using one of the iterative methods, like Newton's method, the secant method, or Brent's method

$$\begin{cases} \eta_1(\omega,\alpha) = 0 \\ \eta_2(\omega,\alpha) = 0 \end{cases} \tag{9}$$

to find the pairs of parameters (ω, α) that satisfy the boundary conditions at the endpoint. When the natural frequencies ω_n and the corresponding initial conditions α_n are known, we can return to differential equation (8) and solve it n times, numerically obtaining n natural modes.

3 Numerical examples

To verify the correctness and effectiveness of the two-parameter single shooting method two examples are taken into account: a non-prismatic simply supported concrete beam and a prismatic propped cantilever steel beam resting on the Winkler foundation and loaded by compressive force. All numerical analyses based on the proposed method were carried out using Mathematica software. The obtained results in the form of natural frequencies were compared with solutions computed by the commercial finite element program SOFiSTiK.

3.1 Non-prismatic beam

The simply supported beam with a span $L = 5.0$ m, a rectangular cross-section of variable width $b(x) = 0.247561 + 0.164911x - 0.00692905x^3$ and a height $h(x) = 0.8 - 0.1x$ (in meters) is made of concrete C 25/30 with a density of $\rho = 2\,400$ kg/m^3 and Young's modulus $E = 31\,476$ MPa. According to the procedure described in the previous point, two hit surfaces were computed by solving Eq. (8) with initial conditions: $f(0) = 0$, $f'(0) = 1$ (normalization), $f''(0) = 0$ and $f'''(0) = \alpha$. The first hit surface represents the values of beam displacements at the end point: $\eta_1(\omega,\alpha) = d^0 f(L) / dx^0 = f(L)$, whereas for the second one the values of the second derivative are related to the bending moment at the end node $\eta_2(\omega,\alpha) = d^2 f(L) / dx^2 = f''(L)$. Both surfaces are plotted in Fig. 1, which also illustrates

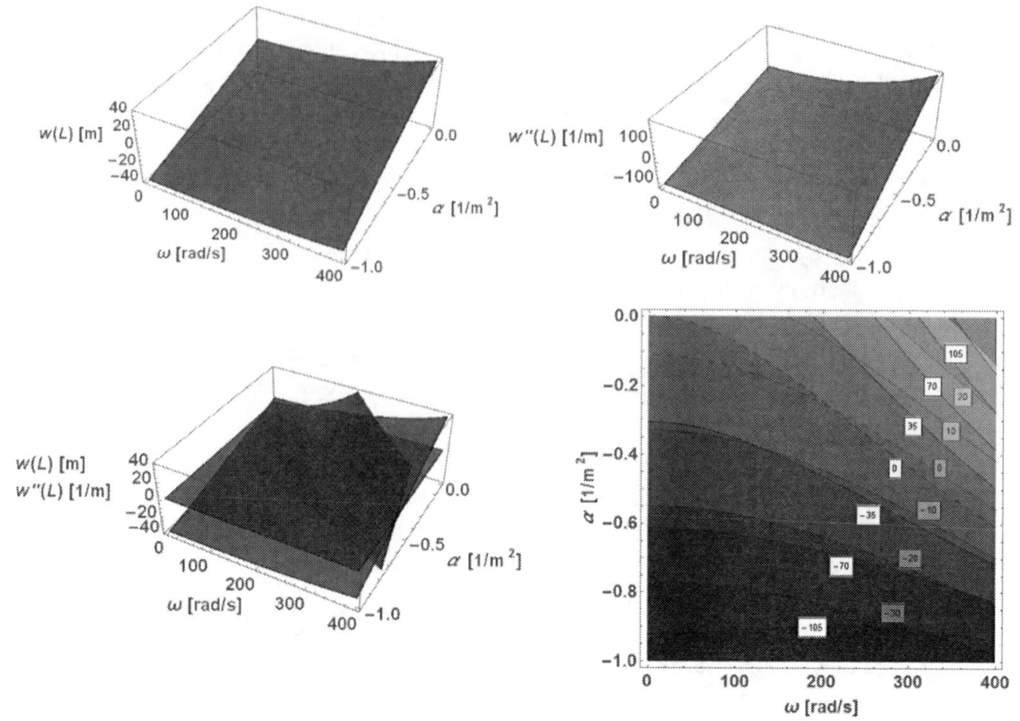

Fig. 1. Hit surfaces obtained by the two-parameter shooting method for the non-prismatic beam.

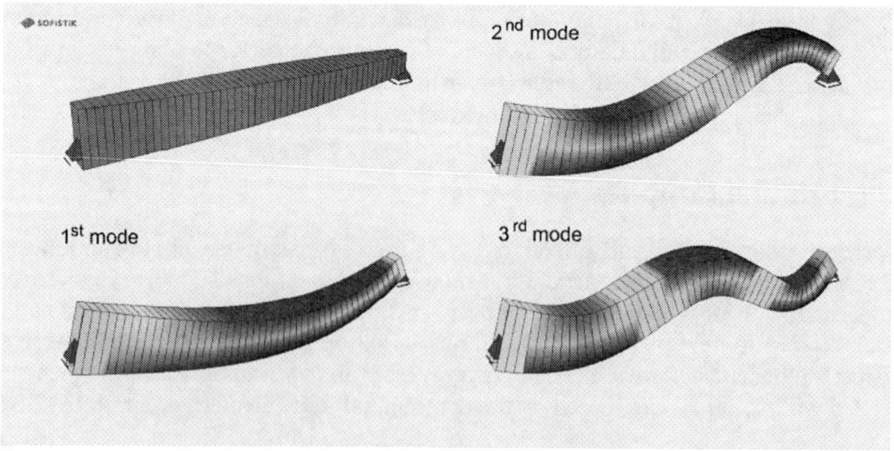

Fig. 2. The non-prismatic concrete beam and three first natural modes calculated in SOFiSTiK.

Table 1. The first three natural frequencies of the non-prismatic beam.

Mode no.	Angular frequency ω [rad/s]		Relative error [%]
	Mathematica	SOFiSTiK	
1	207.50	204.57	1.41
2	856.89	845.61	1.32
3	1933.15	1910.99	1.15

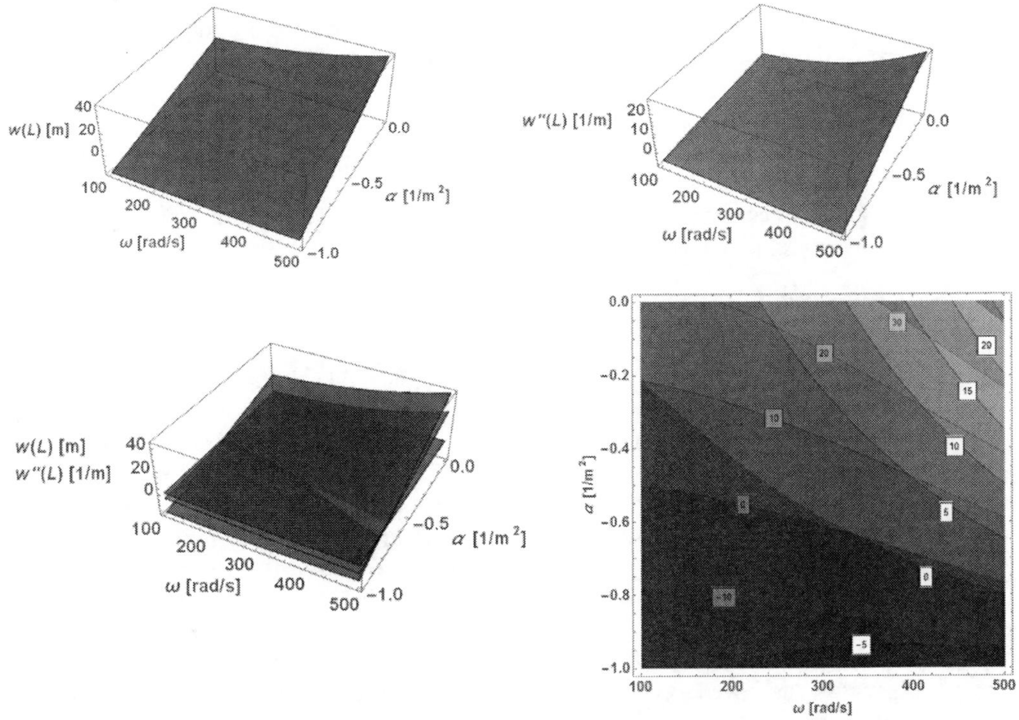

Fig. 3. Hit surfaces obtained by the two-parameter shooting method for the prismatic beam.

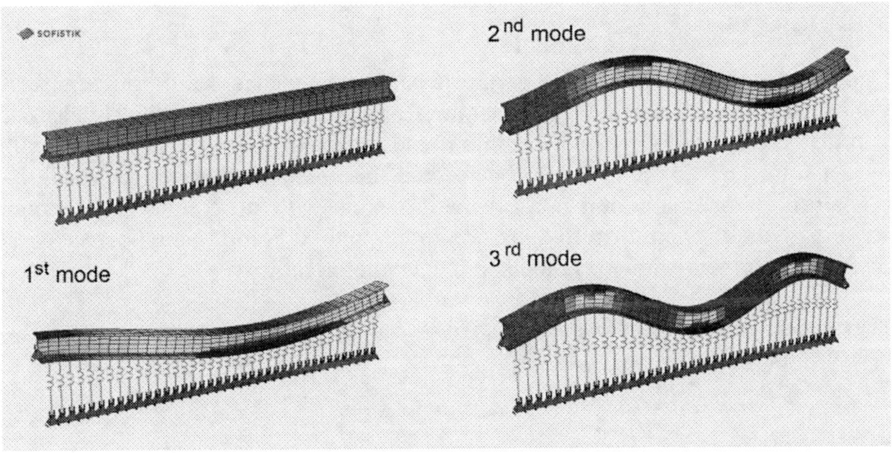

Fig. 4. The prismatic steel beam and three first natural modes calculated in SOFiSTiK.

Table 2. The first three natural frequencies of the prismatic beam.

Mode no.	Angular frequency ω [rad/s]		Relative error [%]
	Mathematica	SOFiSTiK	
1	335.25	335.45	0.06
2	1100.85	1103.30	0.22
3	2306.18	2317.46	0.49

the process of solving the system of equations (9). The first intersection point of the two dashed zero contour lines indicates the value of the first natural frequency of the beam. Further frequencies, not shown in the Fig. 1, are listed in Table 1. To verify the solutions obtained by the two-parameter single shooting method, the same beam was modelled in SOFiSTiK (see Fig. 2). After dividing the structure into 50 finite elements, eigenvalue analysis was performed using the Lanczos method [8]. The computed frequencies and modes are presented in Table 1. As we can see, the results differ slightly, and the maximum relative error does not exceed 1.5%.

3.2 Prismatic beam subjected to axial force

In the second example the propped cantilever double T-beam resting on the Winkler foundation with the stiffness parameter $k = 100$ kN/m^2 is considered. The beam with a rolled HEB 360 cross-section and span $L = 6.0$ m is made of structural steel S 235 with Young's modulus $E = 210\ 000$ MPa and a density of $\rho = 7\ 850$ kg/m^3. The compressive point load $P = 2\ 500$ kN is applied to the end node of the beam. As in the previous example, two hit surfaces (see Fig. 3) were computed by solving Eq. (8) with initial conditions: $f(0) = 0$, $f'(0) = 0$, $f''(0) = 1$ and $f'''(0) = \alpha$. It should be noted that in this case the natural modes are normalized not by the first but the second derivative of the displacement function. The surfaces and their contour plots are presented in Fig. 3. The obtained results are compared in Table 2 with natural frequencies computed by SOFiSTiK (see Fig. 4). A very good agreement is achieved again, which confirms the effectiveness and universality of the proposed method.

4 Conclusions

A novel approach of calculating natural frequencies and modes of single-span beams has been presented. The proposed two-parameter single shooting method is based on the fourth order ordinary differential equation and enables complex problems with two unknown parameters to be solved. Therefore, the method can be successfully used in the analysis of prismatic and non-prismatic beams with different support conditions. The performed analyses confirm that the presented method is formulated correctly and can be used for solving many other problems in a new numerical way.

Calculations were carried out using resources provided by the Wroclaw Centre for Networking and Supercomputing (http://wcss.pl), Grant No. 365.

References

1. D. Zwillinger, *Handbook of Differential Equations* (Academic Press, San Diego, 1997)
2. M. A. El-Gebeily, B. S. Attili, Computers and Mathematics with Applications **45** (2003)
3. P. F. Pai, A. N. Palazotto, Int. J. Solid Structures **33**, 9 (1996)
4. S. N. Ha, Computers and Mathematics with Applications **42** (2001)
5. S. M. Ibrahim, B. P. Patel,Y. Nath, Int. Journal of Non-Linear Mechanics **44** (2009)
6. P. Ribeiro, Computers and Structures **82** (2004)
7. W. Glabisz, K. Jarczewska, W. Zielichowski-Haber, Archives of Civil and Mechanical Engineering **13**, 1 (2013)
8. K. J. Bathe, *Finite Element Procedures* (Prentice-Hall, New Jersey, 1996)

Historical church diagnostic with regards to impact of traffic included vibrations

Krzysztof Kozioł[1,*]

[1]Cracow University of Technology, Institute of Structural Mechanics, 31-155 Kraków, Warszawska 24 St., Poland

Abstract. In the article problem of a comprehensive diagnosis of a historic building with damages is discussed. Dynamic investigations carried out in the church of St. Nicholas in the center of Krakow are an example of diagnostic tests provided to determine the causes of damage. Because the tested structure is exposed to traffic vibrations caused by wheeled vehicles and train passes, special attention has been paid to the impact of these vibrations onto a historic object. Results of measurements of these vibrations as well as dynamic analyzes obtained using FEM model of the investigated church were given.

1 Introduction

A long with the development of road infrastructure, the sources of transport vibrations are brought closer to the existing buildings including historical objects. Diagnostics of the technical condition of these objects cannot ignore the influence of vibrations generated during the passage of vehicles. Appropriate diagnostic analyzes should be carried out taking into account not only the current impact on the building but also occurring in the past.
Dynamic diagnosis regarding the historical church in Krakow [1] may be a good example confirming the above statement. It is therefore worth disseminating the procedure used to develop this diagnosis.

2 Characteristics of the object and the purpose of the diagnosis

Church of St. Nicholas is located in the central part of Krakow. Near the church, from the eastern side, there are located train tracks on the embankment and on the north side - a street on which cars move (see Figure 1).
This church is one of the oldest historic buildings in Krakow. It is under heritage protection. The origins of the building date back to the 12th century. It is not a homogeneous object - it has been rebuilt and expanded several times. In its present architectural form it is a three-nave, Baroque and neo-Baroque basilica. However, Romanesque and Gothic relicts can be found in the walls of the temple.

* Corresponding author: kkoziol@pk.edu.pl

At the end of the 60s of the last century there has been a marked increase in the intensity of movement of trains on tracks located in the vicinity of the church. Dynamic measurements were made (see [2]) and after a detailed analysis it was proposed to use the bulkhead construction (vibroinsulating barrier) in the ground to a depth of 3.0 m as a technical way to reduce vibrations generated and transferred through the ground to the church during train passages. Subsequent decisions that reduced the intensity of railway traffic and redirected freight trains to different line, bypassing the Krakow Main Station, caused the withdrawal of the railway line user from the implementation of the vibroinsulating barrier in the ground.

Fig. 1. Location of the analyzed object with vibration sources.

In the 70's and 90's of the last century reconstruction of the gable wall from the west side and reinforcement and renovation of the church roof were made. Currently, numerous damages in the form of scratches and cracks can be found on the church walls and vaults. Before commencing renovation works it was necessary to determine the causes of these damages and to give conclusions and recommendations regarding the repair and protection of the church.

3 The scope of carried out diagnostic investigations

A comprehensive analysis of the dynamic response of the church structure contained:

- inventory, analysis and evaluation of structural damage,
- vibration measurements at selected points generated during the passage of trains and cars,
- computer analysis of the structure resistance to paraseismic effects caused by the passage of trains and cars,
- outcrops in the ground between the east wall and the base of the railway embankment in order to check the existence of anti-shock protection along the railway track and evaluate its technical condition,
- complete geotechnical documentation of the ground below analyzed object,
- outcrops and checking the technical condition of rainwater drainage installations.

On this basis and taking into account information contained in [3] following was developed:

- guidelines and recommendations for the construction-executive project of reinforcing the ground under the foundations,

- guidelines and recommendations for the construction-executive project of necessary repairs and strengthening the structural elements of the historic church,
- guidelines and recommendations for the scope and range of execution of anti-shock protection under railway tracks related to the planned reconstruction and modernization of tracks.

4 Description of the damages

The analysis of cracks on the surfaces of the walls is usually the most reliable source of information to establish the mechanisms and causes of the damage. On the basis of a detailed inventory of scratches and cracks in the walls of the church structure it was found that they are cross-cut, i.e. they pass through the entire thickness of the walls. In the case of the vaults, identification of the cracks' progress into the structure was not possible due to the lack of direct access. It seems that the location of some cracks in the vaults may indicate their cross-nature, which results in the loss of cooperation between the fragments of the vault separated by such a crack.

On the transverse walls in the southern side nave there are scratches in the vicinity of external buttresses. The nature of these scratches is directly related to the lack of bonds between the transverse walls and the original external church buttresses built in the early twentieth century. The existing cracks reflect the original ledges and faults located on the outer edge of the scarps.

The largest scale of splits and cracks occurs in the south-western corner of the church. Therefore, it was analyzed whether the substructure was loosen under this part of the structure. It turned out that archaeological excavations were performed here in the past. However, no documentation regarding these works has been preserved. Therefore, it is not possible to state clearly what was the depth of excavations and size of the works carried out there. The unknown historical wall discovered then in this region may indicate a wide variety of foundation conditions. It is also known that in the area of the described corner of the church there is a connection to the drainage system of the drainage pipe of the roof. Periodic leaks of rainwater to the ground may cause it to loose as a result of washing away silty parts. Despite significant extent and gap between the cracks at this corner of the church, cracks distribution on the north facade does not confirm the uneven foundations settlement as the dominant cause of damage. Possible settlement could only deepen the scale of damage.

The soil substrate was also tested at the location of the church. On this basis geotechnical conditions of foundation below the church were considered favorable. A review and monitoring of the entire storm sewer system was also carried out. The review concerned gullies inlets starting from inspection grates to collector wells and channels between wells. After checking over 31 linear meters of the installation and carrying out flow tests, investigated installation was confirmed to be unobstructed. However, it was found that the installation is in poor technical condition and needs renovation particularly detected cracks should be repaired.

5 Investigation and evaluation of the influence of communication sources of vibrations onto church structure

Vibration were measured at selected points located on the church structure. These vibrations were generated by passing trains along the Krakow-Medyka route no. 91 and by cars passing along Kopernika st. Acceleration of vibrations components were measured in three directions: two horizontal “x” (along the longitudinal axis of the church –

perpendicular to the track line) and "y" (perpendicular to the longitudinal axis of the church – parallel to the track line) and one vertical direction "z". Measurements were made at three measuring points on the structure at ground level (on the foundation) and respectively at three points at the level of the wall top.

Below are presented maximum values of vibration acceleration recorded during the passage of trains:

- at the ground level (foundation): component "x" - 1,25 cm/s^2, component "y" - 2,12 cm/s^2(vibro-acceleration shown on fig.2) component "z" - 1,34 cm/s^2,
- at the level of the wall's top: component "x" - 4,64cm/s^2, component "y" - 7,73 cm/s^2, component "z" - 4,72 cm/s^2.

the cars:

- at the ground level (foundation): component "x" - 0,22 cm/s^2, component "y" – 0,28 cm/s^2, component "z" – 0,32 cm/s^2,
- at the level of the wall's top: component "x" -1,12 cm/s^2, component "y" – 1,33 cm/s^2, component "z" – 0,77 cm/s^2.

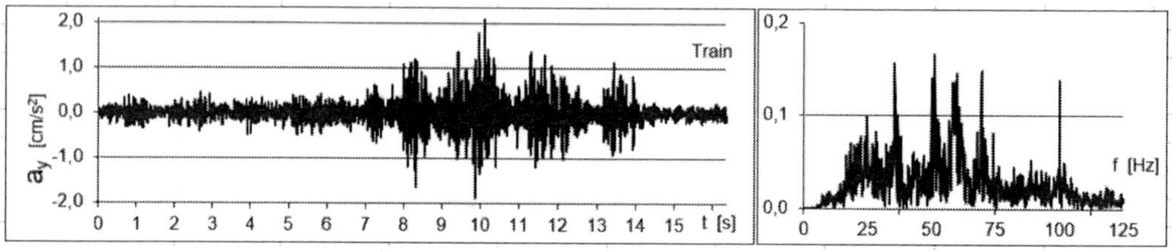

Fig. 2. Vibro-acceleration (and its frequency spectrum) of foundation in the direction of y.

The level of vibrations caused by passing cars along Kopernika st. was several times smaller than the values given above.

Recorded vibrations waveform were used to carry out analysis allowing assessment of the influence of vibrations of communication origin onto the church structure. For this purpose a 3D FEM model of the structure was prepared. Fig. 3 shows the applied division of the structure into finite elements.

Fig. 3. Division of the computational model into finite elements and photo google maps.

Dimensions and distribution of finite elements were adopted in accordance with the available documentation, including its verification during the damage inventory. Applied calculation model also takes into account the larger damage occurring in the construction, assuming appropriate density of the element mesh in the vicinity of these locations (see Fig. 4).

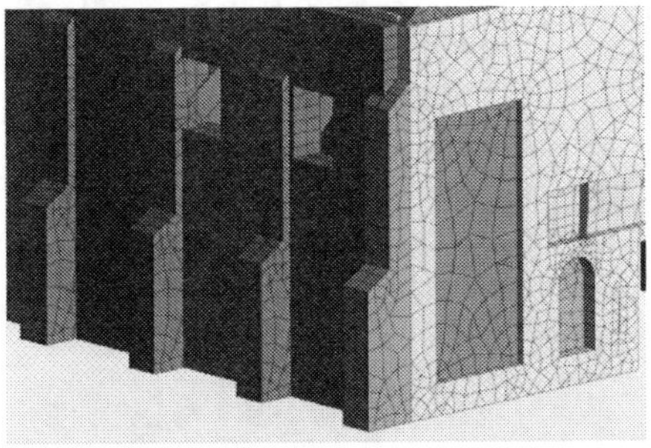

Fig. 4. Part of a FEM model with enlarged mesh density around visible damages.

During FEM model preparation, it was assumed that the materials present in the structure are characterized by the following values:

- brick wall: Young's modulus E = 3,2 GPa; Poisson's ratio ν= 0,2; mass density ρ=2000 kg/m3, material is present in the walls and foundations;
- pine wood: Young's modulus E = 6,0 GPa, shear modulus G=0,50 GPa, mass density ρ= 850 kg/m3, material is present in the roof structure.

From recorded measurements of church vibrations for further analysis measurements were chosen as containing the most unfavorable values. Vibrograms registered directly on the foundation of the church were established as a kinematic excitation used in the computational model. It was assumed that this excitation occurs at each node at the interface of the foundation with the ground. The level of the wall's top was taken as the reference. Therefore, the acceleration obtained from the numerical model in the level of the wall's top was compared with the vibration measurements registered during the same measurements. Sufficient results were obtained.

The subject of further analysis was the usage of the FEM model to check if there is a cause-and-effect relationship between the excitation generated by the train passage and the cracks onto object. The tensile stresses occurring in the object were analyzed: in the theoretical cross sections not scratched and real cross sections cracked. An exemplary map of stresses occurring on the south and west facades is shown in Fig. 5a, 5b.

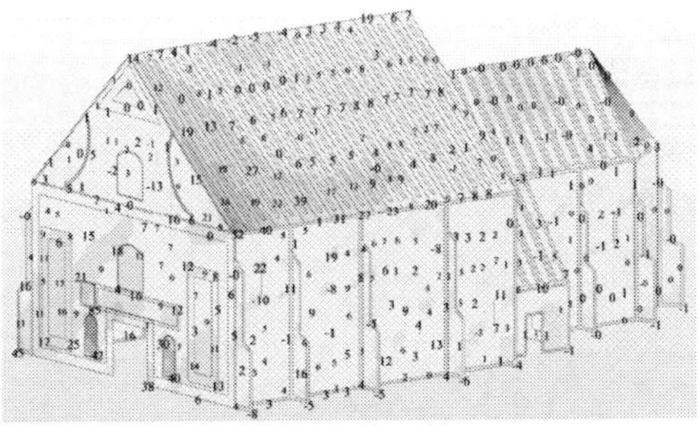

Fig. 5a. An exemplary map of the maximum main stresses (values in Pa, should be treated as variable sign) - caused by vibrations generated by train passage.

Fig. 5b. An exemplary map of the maximum main stresses (values in Pa, should be treated as variable sign) - caused by vibrations generated by car.

Basing on the calculations, it was found that the analyzed dynamic action induces tensile stresses with the following values: in the vaults they did not exceed 22 Pa, in longitudinal walls (sideways, perpendicular to the railway track) they reached 45 Pa and in the presbytery and in the front wall they did not exceed 43 Pa, with the exception of the window opening in the front wall, where they reached the value of 85 Pa.

The stress values obtained from calculations do not indicate that the vibrations caused by currently occurring trains and cars passes could be primary cause of damage to the structural elements of the church. However, they can lead to the disclosure and development of damage caused earlier by different causes.

6 Conclusions

Relatively extensive investigations and analysis have identified condition of the structure and outline a course of action with the technical lead for the repair and strengthening. The group of main recommendations includes:

- indication that there is no need to strengthen the ground under foundations,
- indication - in the current traffic and rails conditions – as the most appropriate solution causing reduction of vibrations transmitted to the structure by applying vibration isolation in the construction of railway tracks, which are passing in the vicinity of the church (within the section about 150m – 75 m north and south from the presbytery of the church),
- the level of stress caused by the loads caused by the passage of trains and cars showed that they do not affect the safety of the structure
- indication of the need to repair damage observed in the walls and provision of the protection works scope.

References

1. K. Stypuła. (head of the authors' team), Specialist structural expertise with making vibration measurements and dynamic computer analysis as well as geotechnical investigations of the ground foundation of the St. Nicholas church at Kopernika 9 st in Cracow, Cracow University of Technology, Institute of Structural Mechanics, November 2009
2. K. Kozioł, K. Stypuła, Simulation calculations during design of vibration isolation of rail surfaces. Selected applications (in Polish), Land – Air – Water ROUTES, No. **10** (2010)
3. PN-B-02170:2016 Evaluation of the harmfulness of building vibrations due to ground motion (in Polish)

MATEC Web of Conferences **285**, 00007 (2019) https://doi.org/10.1051/matecconf/201928500007
DYNKON 2019

Reduction of vibration in railway surfaces

Krzysztof Kozioł[1,*]

[1]Cracow University of Technology, Institute of Structural Mechanics, 31-155 Kraków, Warszawska 24 St., Poland

Abstract. Presented paper concerns the problem of reduction of the impact of vibrations generated by the metro on neighboring buildings and people staying inside. The issue of acceptable levels of these impacts in accordance with Polish standards and ISO (International Standard Organization) has been discussed. Basing on the example of research carried out by the author in the Warsaw Metro and his experience acquired during the design of vibroisolations of the final section of the first subway line in Warsaw, methods of testing dynamic interactions and predicting their level using FEM and simulation calculations on models were presented

1 Impact of metro traffic on buildings and people in buildings

Main impact of the subway on the environment is related with emission of vibrations generated by the passage of trains in the tunnel. These vibrations spread in the soil (as a result of wave propagation) and affect neighboring buildings and people in these buildings (Fig. 1).

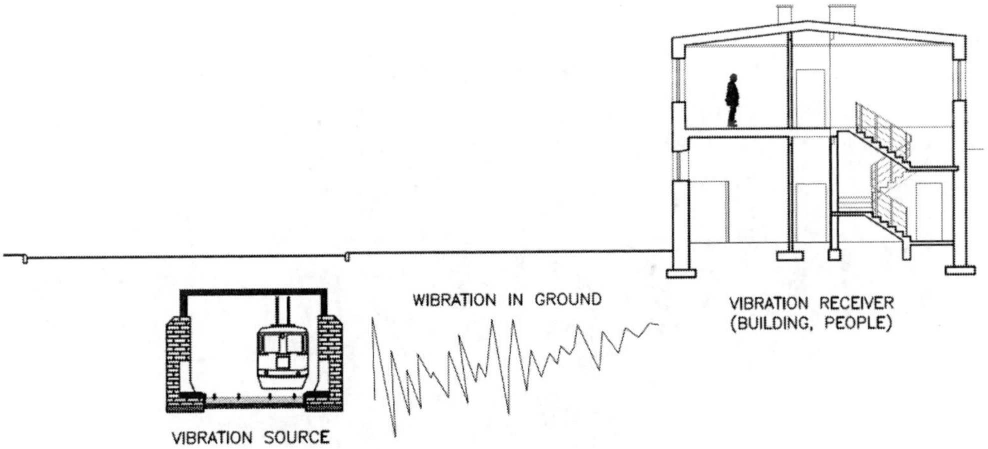

Fig. 1. Diagram of vibration emission from the subway tunnel.

* Corresponding author: kkoziol@pk.edu.pl

The best way to avoid mentioned nuisance associated with these influences is to reduce the vibrations emitted from the subway tunnel by choosing the appropriate structure of the rail surface with usage of vibroinsulating elements. Well-designed vibration isolation of the rail surface gives a significant reduction of the level of vibrations emitted to the environment. Below are presented scientific-research issues regarding the proper selection of the rail surface structure due to limitation of mentioned vibration emission so that the permissible vibration levels specified in the relevant standards are not exceeded [1-4].

2 Acceptable levels of interactions according to Polish and ISO standards requirements

The legal basis and detailed regulations related to assessments of the impact of vibrations on buildings and people in buildings are contained in two Polish standards PN-B-02170:2016 - Assessment of harmfulness of vibration transmitted by the ground on buildings [3] and PN-B-02171:2017 - Assessment of the impact of vibrations on people in buildings [4].

Vibrations measured at appropriate points of the structure or obtained as a result of numeric calculations based on input data from such measurements are assessed. Generally measurements concern three components of vibrations: two mutually perpendicular horizontal directions marked "x" and "y" and a vertical component "z".

The PN-B-02170:2016 standard provides two ways of assessing the impact of vibrations onto building structure:

- full assessment - applicable for each type of building,
- approximate assessment (using scales of dynamic impact SWD-I and SWD-II) - applicable only two specific building classes.

Full assessment of the impact of vibrations on the building structure should be made in accordance with principles of building dynamics calculations. This requires preparing a FEM model of the building that could be used for dynamic calculations (Figure 2) and forcing its vibrations. This could be done either by applying vibrations to the foundations - measured or predicted.

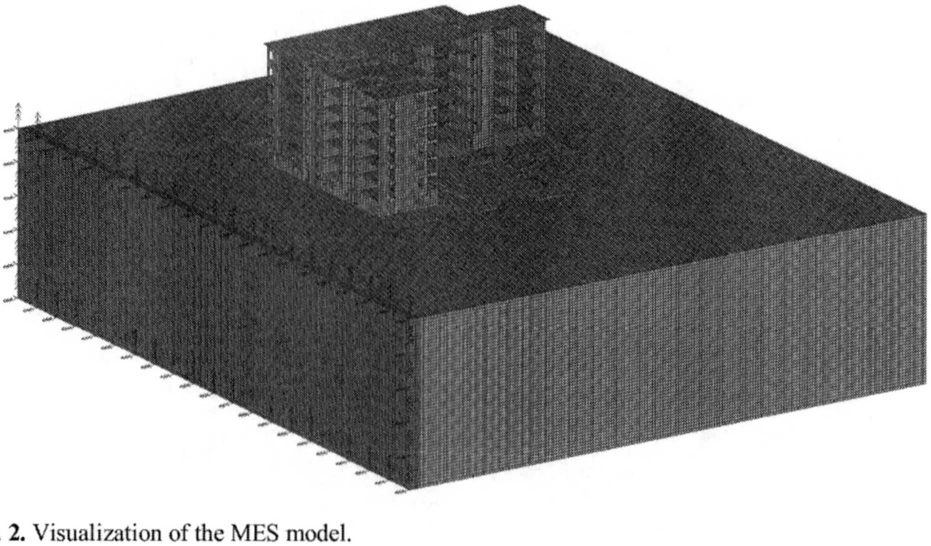

Fig. 2. Visualization of the MES model.

Mentioned assessment consists in determination of inertia forces acting additionally on the structure during vibrations, loading of the FEM model with all static and dynamic loads and checking the strength of such loaded structure in accordance with requirements of existing standards for static calculations and structure design
Taking into account labor consumption of the calculations discussed above and the fact that horizontal vibrations have a decisive influence onto structure of buildings, the approximate way of assessing dynamic influences (precisely influence of horizontal vibration components) on the building using of so-called dynamic influence scale: SWD-I and SWD-II were introduced. These scales concern buildings made of masonry elements (intended for manual laying such as bricks, hollow bricks, etc.) and prefabricated using large-blocks and large-panels.
The SWD-I scale [4] refers to buildings with a compact shape, small external dimensions of a plan projection (not exceeding 15 m), one or two-story and with a height not exceeding any of the plan projection dimensions.
The SWD-II scale [4] refers to buildings no higher than five floors, which height is smaller than the double smallest building width in a plan projection and to low buildings (up to two floors) but not fulfilling conditions specified for the SWD-I scale.
Evaluation consists of horizontal components of the building vibrations, measured in a rigid node of the structure (from the source of vibrations) in the ground level or on the foundation of the building. Recorded vibrations are analyzed in order to determine the maximum peak values of displacements or accelerations amplitudes (vertical axis of the scale) in individual frequency bands (horizontal axis of the scale). Based on the results of the analyzes, buildings vibrations could be qualified o one of five harmful zones:

- zone I – vibrations imperceptible for the building;
- zone II – vibrations perceptible by the building, but not harmful to its construction;
- zone III – vibrations harmful to the building, causing local scratches and cracks;
- zone IV – vibrations of high harmfulness to the building, posing a threat to human safety;
- zone V – vibrations causing building failure due to wall collapse, falling ceilings, etc. The building cannot be used.

Fig. 3 presents exemplary results of chosen building vibration analysis in 1/3 octave bands in the case of the assessment of the impact of these vibrations using SWD-I and SWD-II scales - vibrations are imperceptible for the building.

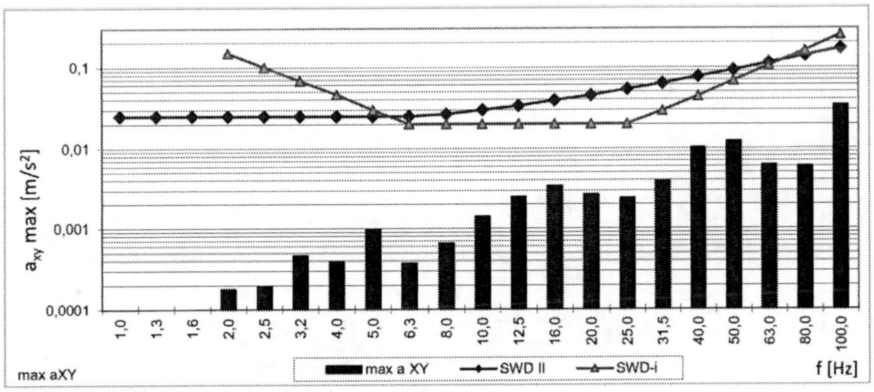

Fig. 3. An exemplary assessment of the impact of vibrations on a building using the SWD-II scale.

Evaluation of the impact of vibrations on people is performed in Poland on the basis of PN/B-02171:2017 and international standards : ISO 2631-2 [1] and ISO 10137 [2].. This standard defines the thresholds of perceived vibrations by people and allowable values of mechanical vibration parameters to ensure the required comfort of people staying inside building, depending on:

- purpose of the room inside the building (residential, offices, workshops, hospitals, precise laboratories, etc.);
- time of the day when vibrations occur (day i.e. from 6 am to 2 pm or night, i.e. from 8 pm to 6 pm);
- nature and repeatability of vibrations;
- direction of vibrations (horizontal or vertical) and position of human during reception of vibrations (standing or lying).

These values concern vibrations in the vertical direction "z" (along the axis of human spine) and vibrations in the "xy" directions (horizontal vibrations, transverse directions to the axis of the spine). Described assessment is based on the results of frequency analysis of vibrations recorded at the place of their reception by human. As a result of such analysis effective values of accelerations (or velocities) vibrations so-called RMS (root mean square) in 1/3 octave (terce) frequency bands are obtained and compared with the permissible values of these parameters. Permissible values are determined by multiplying the vibration perceptibility threshold value by the multiplier n given in table 1.

Table 1. Values of the "n" coefficient according to PN-B-02171:2017 [4]

Purpose of the room	Time of the day	Value of the coefficient during vibrations:	
		Fixed (continuous or intermittent) with a multiplicity greater than 10 per day	Occasional
Hospitals (operating rooms), precise laboratories	Day	1	1
	Night		
Hospitals (patients rooms)	Day	2	8
	Night	1	4
Apartments	Day	4	32
	Night	1,4	4
Offices. schools	Day	4	64
	Night		
Workshops	Day	8	128
	Night		

The comfort limits as a function of vibration acceleration are shown in Fig. 4.

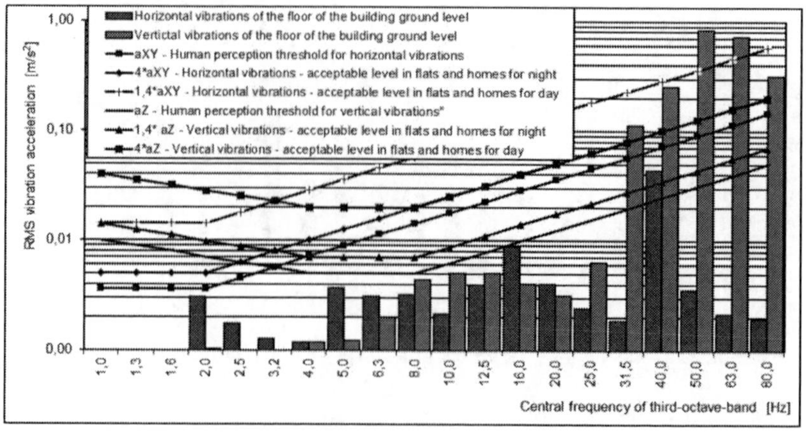

Fig. 4. Impact of vibrations on people in homes.

Analysis results of the impact of vibrations on humans can be briefly expressed using the WODL(f) index - human perception of vibrations index. This ratio expresses the highest value of the ratio of the effective value of acceleration (or velocity) determined due to analysis of vibrations (obtained from measurement or calculation) – RMSa(f) to the effective acceleration (or velocity) value corresponding to the human vibration perceptibility threshold in the same 1/3 octave (terce) frequency band. The value of the WODL index is given together with the information about the central frequency of the ⅓ octave band in which WODL is determined – min((aXY(f),aZ(f))

$$\text{WODL}(f)=\max(\text{RMSa}(f)\,/\,\min((\text{aXY}(f),\text{aZ}(f)) \quad (1)$$

Many times during analysis of impact of vibrations on people, frequency is omitted giving only the maximum value of the WODL coefficient. As a result one loses information about the frequency at which the threshold of perceived vibrations is exceeded.

In an analogous manner, in buildings where the impact of vibrations is assessed using the SWD scales, one can determine the building vibration perception index - WODB. It is defined as the maximum value of the ratio of the peak value of acceleration (or displacement) to the acceleration (or displacement) value corresponding to the vibration perceptibility limit of the building in the same 1/3 octave (terce) frequency band (specifying the center frequency of this band).

3 Methodology of investigating dynamic interactions and forecasting their level on the example of research conducted in the Warsaw Metro

Based on the author's many years of experience related to studies of vibrations induced by metro operation as well as analysis of the impact of these vibrations on neighboring buildings and people inside buildings [5] a scheme was developed to account for these impacts in the case of the construction of rail structure on new subway sections (Fig. 5). This scheme can also be used in the case of modernization of the track structure (metro, tram, etc.).

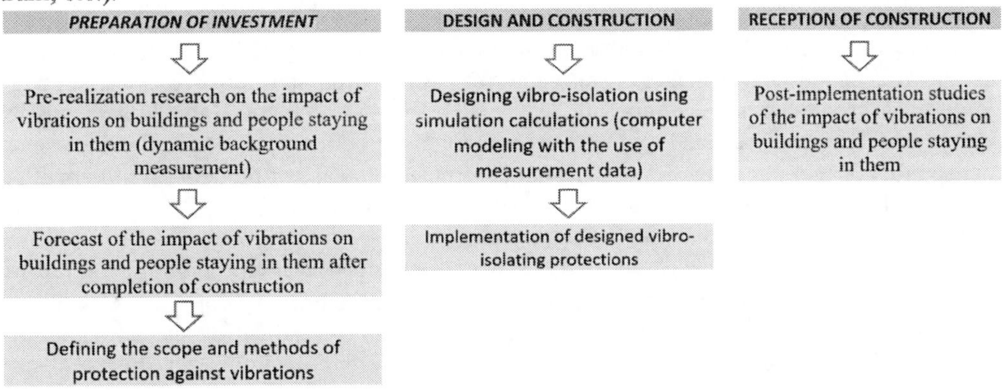

Fig. 5. The scheme of taking into account environmental impact of vibrations in case of construction of a subway rail.

Procedures described above have been successfully used for a long time during the construction of subsequent metro sections in Warsaw. As an example, it is possible to provide activities undertaken on the final section of the first metro line in Warsaw, comprising three stations and three track tunnels (approx. 3 km). Within the dynamic impacts of the subway zone, whose range was in Warsaw (shallow subway) was set at 40 m in a plane projection from the outer walls of the tunnel, there were 129 buildings, of which 36 were selected as a representative structures in order to perform dynamic background measurements and vibration impact predictions. Earlier in all buildings, an inventory of the technical condition of their structure was carried out (along with photographic documentation of damages), and then in representative buildings the following activities were taken: dynamic background measurements and prediction of impact of vibrations onto building and people staying in the building. Because in a significant number of buildings the predicted level of vibration impact on people was too high (significant exceeding of the human vibration threshold) it was necessary to design in the track tunnel (fig. 6) at segments where these buildings are located, the new non ballasted track using the EBS system (Embedded Block System) produced by Edilon and, where it was necessary, also anti-vibration mats (fig. 7).

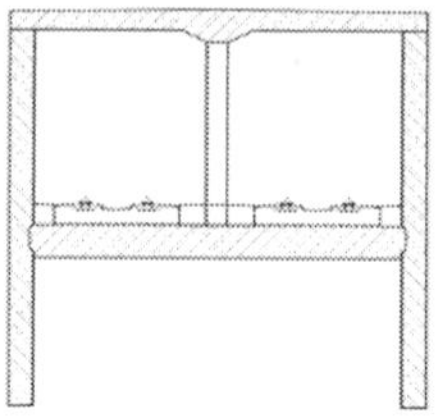

Fig. 6. Schematic cross section of the Warsaw metro tunnel within the final section of the first Warsaw metro line.

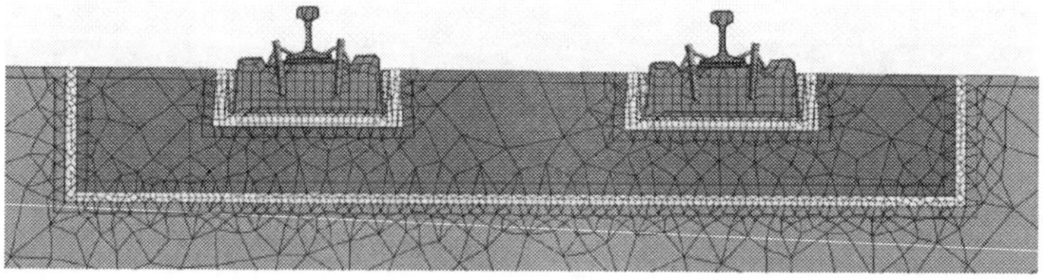

Fig. 7. Diagram of the subway rail details with EBS Edilon and vibration isolation mat.

Computational models (FEM) for both analysed cases were constructed; material parameters of individual elements were determined based on laboratory test or assumed as suggested in standards, remembering (especially for soil elements) about their possible deviations. Presented results include values obtained for boundary and mean values (in the sense of the arithmetic average) of material parameters. In case of models extended with an additional vibration insulation layer, a mat with a thickness of 2 and 2.5 cm and with a dynamic stiffness (depending on frequency and pressure) ranging from 0.014 to 0.02 N / mm3 was used. In the presented analysis it was assumed that the damping of the mat expressed by the loss angle tg(τ) is equal to 0.3. Onto the vibration propagation path, there are places where a special approach requiring experimental verification should be applied. These include the point of contact between two different mediums. In this case, it is required to perform experimental tests that will verify the parameters applied at the interface contact. An exemplary function of the transition between the ground and the building is described in [6-8] and in Figure 8. As can be seen from Fig. 10, the amplitudes characterising lower frequencies (both horizontal and vertical) can increase during transition from the ground to the building.

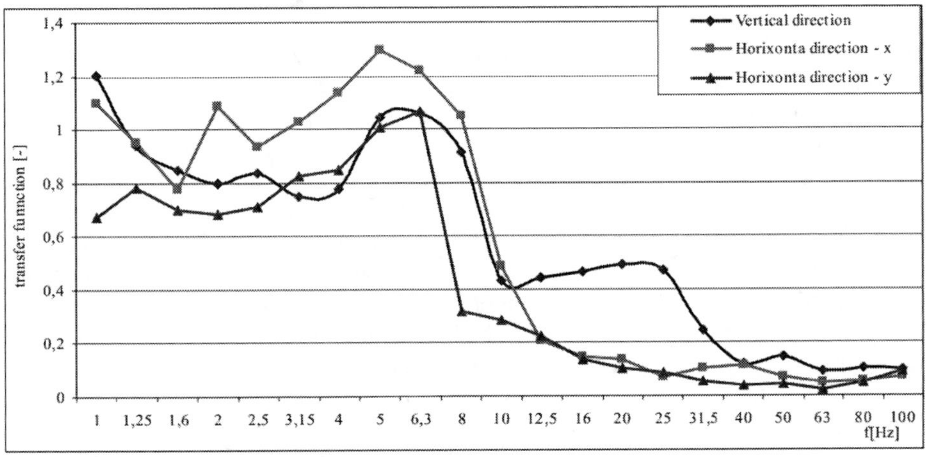

Fig 8. Transfer function of the acceleration from the ground to building for the vertical direction z and horizontal directions x, y [8].

At the beginning sense of usage of such a rail structure was analyzed. For this purpose, analyzes of the human perception of vibrations index (WODL) were made in one of investigated building namely most exposed to vibrations, assuming different variants of vibration isolation or lack thereof. An exemplary analysis is presented in Fig. 6. Selected results of these analyzes are presented in Table 2, showing the most unfavorable values of the human perception of vibrations index (WODL) obtained from described analyzes.

Table 2. List of the most unfavorable results of simulation calculations of the impact of vibrations on people in the K61 building at different variants of the track structure.

Construction variant	WODL index	Comments
Existing typical construction of the metro rail structure in the route tunnel **without any vibro-isolation**	13,9	Inadequate protection
Structure of the rail track in the route tunnel with EBS, bottom **plate 60 cm thick, without vibro-insulating mat**	4,7	Inadequate protection
Structure of the rail track in the route tunnel with EBS, bottom **plate 60 cm thick, with 2.0 cm thick mat**	1,7	Inadequate protection
Structure of the rail track in the route tunnel with EBS, bottom **plate thickened up to 70 cm, with 2.0 cm thick mat**	1,1	Inadequate protection
Structure of the rail track in the route tunnel with EBS, bottom **plate thickened up to 80 cm, with 2.5 cm thick mat**	**0,7**	Adequate protection

Conducted analysis shows that usage of the EBS system alone in the analyzed building reduces the WODL index by as much as 66%. Despite such a significant reduction f the impact of vibrations on people, only the combined use of EBS and vibroinsulating mat (with thickening the bottom plate up to 80 cm) reduced the value of this index to the required level (below 1.0).

4 Summary

Impact of vibrations on neighboring buildings and particularly on the residents of these buildings is currently the main problem in the field of environmental protection in the case of subway operation. Primary method of reducing the level of vibration generated by subway is usage of isolation in the rail surface. Usage of an appropriate vibroisolation solution in the construction of a rail surface should be result of designing process of vibroisolation using simulation calculations. The necessity to make calculations confirming the achievement of the required effectiveness of vibroisolation (lowering the level of impact of vibrations on people below the level of perceptibility of these vibrations) in the case of its specific application results from the dependence of vibroisolation effects on many factors e.g.: damping properties of vibroinsulating material, dimensions of used elements, e.g. thickness of vibroisolation mats (taking into account surface deformation), rail stiffness, spacing of supports (supports or block supports) and resulting rail deflections and consequently, changes in excitation characteristics, structure and dimensions (mass) of sub-floor slabs, foundation rigidity, tunnel casing construction, ground and water conditions during propagation from tunnel to building, structure and dynamic features of the building receiving vibrations. As described in the above article in the case of appropriately designed two-stage vibration isolation consisting of the EBS system in combination with usage well-designed (good in terms of dynamic features) vibration isolating mat one can achieve high efficiency, reducing the impact on people in buildings below the threshold of perceptibility vibration by humans.

References

1. ISO 2631-2; Guide to the evaluation of human exposure to whole body vibration. Part 2 – *Vibration in buildings*, (2003)
2. ISO 10137, Bases for design of structures – Serviceability of buildings and walkways against vibrations, (2007)
3. PN-B-02170:2016 Evaluation of the harmfulness of building vibrations due to ground motion (in Polish)
4. PN-B-02171:2017 Evaluation of vibration influence on people in buildings (in Polish).
5. Kozioł K. Influence of different structural solutions for dynamic response of the modernized building. IOP Conference Series: Materials Science and Engineering –, Vol. **245**,dings (in Polish) (2017)
6. Kozioł K. Influence of excitations originating from moving subway onto transferring vibrations from ground to structure. E3S Web of Conferences [online]. – 2016, Vol. **10**, 1st International Conference on the Sustainable Energy and Environment Development SEED 2016, Krakow, Poland, (2016)
7. Kozioł K The influence of vibration propagation from trams on building vibration caused by ground shaking. Proceedings of the Fifteenth International Conference on Civil, Structural and Environmental Engineering Computing, [Prague - Czech Republic, 1-4 September 2015] [USB] / ed. by J. Kruis, Y. Tsompanakis, B. H. V. Topping. – Stirlingshire, UK : Civil-Comp Press, (2015)
8. Kozioł K., „A comparison of the vibration effects caused by city transportation systems on the occupants of buildings" , CCMM Prague-Czech Republic, (Sept. 2015)

Optimal positions of translational vibration absorbers in vibroisolation of mechanical press

Waldemar Łatas*

Institute of Applied Mechanics, Cracow University of Technology,
Al. Jana Pawła II 37, 31-864 Kraków, Poland

Abstract. The paper deals with the problem of optimizing the positions of dynamic vibration absorbers in a mechanical press subjected to harmonic excitation force. Assuming flat motion of a rigid body and small vibrations, a discrete linear dynamic model of the press with absorbers is constructed. Using the results of numerical calculations, the optimal positions of the single or two vibration absorbers tuned to the excitation frequency, for different direction of excitation force, are obtained.

1 Introduction

Tuned mass dampers or dynamic vibration absorbers (i.e. TMDs with no damping) are simple and effective vibration suppressing devices, employed in a variety of problems of different scale and frequency range. In civil engineering TMDs are used in structures with low internal damping and low natural frequencies, such as suspension and cable-stayed bridges, high-rise buildings, masts, chimneys and wind power towers, exposed to wind or earthquake excitations [1-4]. In road and railway bridges, as well as foot bridges, where tuned mass dampers are also widely used, the excitation forces come from vehicle and pedestrian traffic [5-6]. On a smaller case TMDs are efficient devices in industry for vibration protection of equipment, e.g. for high-speed machining chatter suppression [7]. Many theoretical works were devoted to optimization of the tuned mass dampers, of passive, active or semi-active types, for different types of excitations, in both linear and non-linear problems [8-11].

During the operation of machinery such as vibrating screens, mills, textile machines, fans, or presses, occurs dynamic forces transmitted to the ground [12], causing undesirable vibrations and noise in the work area and in the surrounding environment (so called *structure borne* noise and vibrations). Large presses can generate vibrations of frequencies which may coincide with natural frequencies of the floors or walls. The most important issue is the machine isolation by the proper choice of elastic and damping support parameters. If the isolation is not sufficient, it is necessary to use additional means, for example passive tuned mass dampers may be helpful due to the cost and reliability issues. In such situations determining the optimal positions of dampers, which may be limited by technical consideration, is an additional task.

* Corresponding author: latas@mech.pk.edu.pl

The paper presents the problem of vibroisolation of a mechanical press, subjected to inertial excitation due to the reciprocating motion of machine components. Such excitation commonly is of a polyharmonic type, due to the periodic movements of slides along the guides in opposite directions, resulted from the rotation of appropriately shaped cams. Vibrations of the presses in a hall are transmitted to the ground and may cause structure borne vibrations in the surroundings, for example the floors in the adjacent buildings.

If the excitation force spectrum contains significant components of very low frequencies, it may not be possible to use the supercritical vibroisolation (i.e. supports with very low stiffness), even with the pneumatic springs, due to the economical or technical reasons. Then, the used damping pads (often made of elastomers) may amplify some of the harmonics of the excitation force, resulting in exceeding the admissible levels in office rooms due to the impact of vibrations on the human body.

The aim of the paper is to determine the optimal positions and tuning of translational dynamic vibration absorbers (DVA, i.e. the TMDs with no damping device) suppressing the selected harmonics of the force transmitted to the ground. The use of the resonant devices is justified due to the stable working conditions, because drive shaft of the press always rotates at a constant angular velocity.

In numerical calculations, one absorber (horizontal or vertical) and a system of two absorbers (horizontal and vertical) are considered. Because the harmonics relevant to the accuracy of the technological process may vary from those harmonics that are important due to the impact on the human body, a wide frequency range including the all natural frequencies of the system is taken into account.

2 Description of the system

Figure 1 presents the diagram of press resting on four supports, excited by the resultant inertial force $F(t)$. The mass of the press equals $M = 11000\,\mathrm{kg}$, its moment of inertia about the axis passing through the center of mass equals $J_C = 4020\,\mathrm{kgm}^2$, the geometric dimensions are: $b = 0.75$ m, $a = 0.90$ m, the angle of inclination of the force is $\beta_0 = 45^0$.

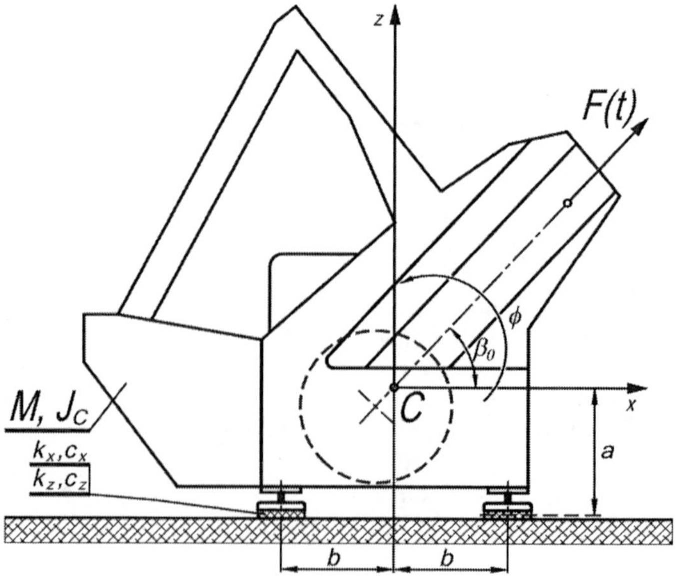

Fig. 1. Press diagram without attached tuned mass dampers.

The horizontal and vertical stiffness and the respective damping coefficients of the supports made of the elastomer pads are as follows: $k_x = 760 \cdot 10^5$ N/m, $c_x = 280 \cdot 10^3$ Ns/m, $k_z = 670 \cdot 10^6$ N/m, $c_z = 540 \cdot 10^3$ Ns/m. The dimensionless damping coefficients in the horizontal and vertical directions are: $\zeta_x = 0.31$, $\zeta_z = 0.20$. The three natural frequencies of the press are approximately (when ignoring damping): $f_1 = 24$ Hz, $f_2 = 78$ Hz, $f_3 = 106$ Hz.

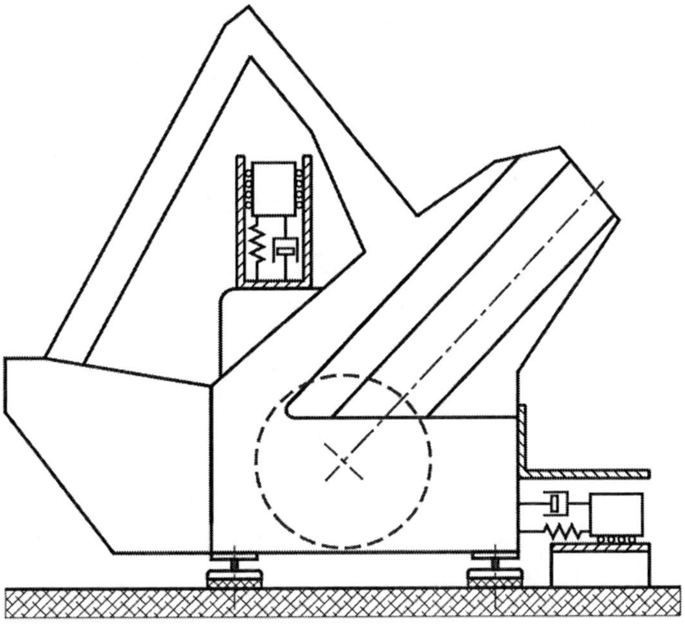

Fig. 2. Press with attached the horizontal and the vertical tuned mass dampers.

Figure 2 shows the press with two attached mass dampers, the vertical moving along the slide attached to the machine and the horizontal moving along the slide fixed to the ground. These dampers move in different ways, but with some simplifying assumptions (mainly neglecting the nonlinear terms) both can be described with the same theoretical model.

The mass dampers have a simple principle of operation, but their effectiveness, in addition to the proper selection of physical parameters, depends significantly on the appropriate location. The best location may not be obvious when the vibrations in different directions are coupled.

3 Model of the press with attached tuned mass dampers

The computational model presented, of a linear type, was built with the assumptions of small vibrations, what is commonly accepted in the vibroisolation problems.

Figure 3 presents a press diagram with a system of two horizontal and two vertical mass dampers attached. The number of mass dampers could be higher, but it is not justified in the considered problem, where the aim is to find the simple and sufficient dampers configuration. Relating to the differences between the dampers mountings shown in Fig. 2, it is assumed that all dampers move along the slides fixed to the ground, without investigating the problems of technical nature.

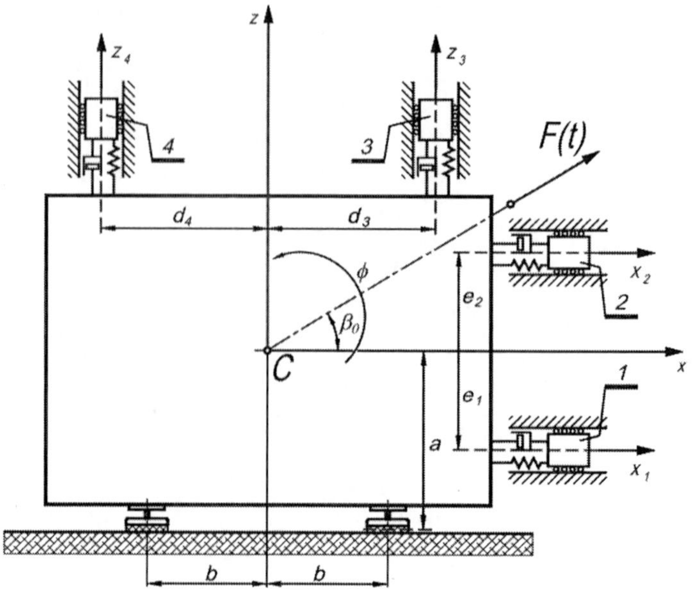

Fig. 3. Schematic arrangement of the horizontal and vertical tuned mass dampers.

In Fig. 3 the dampers are placed outside of the contour of the machine, but they can be moved to the inside of the contour along the motion directions. Additionally, in order to avoid rotating of the press with respect to the vertical axis, one damper in the diagram represents in practice two symmetrically mounted dampers (with its parameters decreased twice).

A flat motion of the system is assumed, described by the following coordinates: x_C, z_C – press mass center displacements in the horizontal and vertical direction, φ – press rotation angle; x_1, x_2 – displacements of the horizontal dampers; z_1, z_2 – displacements of the vertical dampers.

The vibrations of the press with the system of mass dampers attached (shown in Fig. 3) are described by the system of linear differential equations:

$$\begin{aligned}M\ddot{x}_C+(4c_x+c_1+c_2)\dot{x}_C+(4c_xa+c_1e_1-c_2e_2)\dot{\varphi}-c_1\dot{x}_1-c_2\dot{x}_2+(4k_x+k_1+k_2)x_C\\+(4k_xa+k_1e_1-k_2e_2)\varphi-k_1x_1-k_2x_2=F(t)\cos\beta_0\end{aligned}\tag{1}$$

$$\begin{aligned}M\ddot{z}_C+(4c_z+c_3+c_4)\dot{z}_C+(c_3d_3-c_4d_4)\dot{\varphi}-c_3\dot{z}_3-c_4\dot{z}_4\\+(4k_z+k_3+k_4)z_C+(k_3d_3-k_4d_4)\varphi-k_3z_3-k_4z_4=F(t)\sin\beta_0\end{aligned}\tag{2}$$

$$\begin{aligned}J_C\ddot{\varphi}+(4c_xa+c_1e_1-c_2e_2)\dot{x}_C+(c_3d_3-c_4d_4)\dot{z}_C-c_1e_1\dot{x}_1+c_2e_2\dot{x}_2-c_3d_3\dot{z}_3+c_4d_4\dot{z}_4\\+(4c_xa^2+4c_zb^2+c_1e_1^2+c_2e_2^2+c_3d_3^2+c_4d_4^2)\dot{\varphi}-k_1e_1x_1+k_2e_2x_2-k_3d_3z_3+k_4d_4z_4\\+(4k_xa+k_1e_1-k_2e_2)x_C+(k_3d_3-k_4d_4)z_C+(4k_xa^2+4k_zb^2+k_1e_1^2+k_2e_2^2+k_3d_3^2+k_4d_4^2)\varphi=0\end{aligned}\tag{3}$$

$$m_1\ddot{x}_1+c_1\dot{x}_1-c_1\dot{x}_C-c_1e_1\dot{\varphi}+k_1x_1-k_1x_C-k_1e_1\varphi=0\tag{4}$$

$$m_2\ddot{x}_2+c_2\dot{x}_2-c_2\dot{x}_C+c_2e_2\dot{\varphi}+k_2x_2-k_2x_C+k_2e_2\varphi=0\tag{5}$$

$$m_3\ddot{z}_3+c_3\dot{z}_3-c_3\dot{z}_C-c_3d_3\dot{\varphi}+k_3z_3-k_3z_C-k_3d_3\varphi=0\tag{6}$$

$$m_4\ddot{z}_4+c_4\dot{z}_4-c_4\dot{z}_C+c_4d_4\dot{\varphi}+k_4z_4-k_4z_C+k_4d_4\varphi=0\tag{7}$$

In the above equations: m_i, k_i and c_i denote the mass, stiffness and damping coefficient of the *i*-th damper, respectively.

4 Results of numerical calculations

The press drive shaft operates at a constant angular velocity (except for acceleration and braking), so the use of the resonant devices is justified, although they are more sensitive to inaccurate tuning from devices with damping appliances. The possibility of effective vibration attenuation using the single horizontal absorber, the single vertical absorber and the system of two absorbers (one horizontal and one vertical) will be examined.

In the calculations it was assumed the harmonic excitation force: $F(t) = F_0 \sin \omega t$, where $F_0 = const$. The system of differential equations (1)-(7) was solved using the prediction method. In order to determine the necessary coefficients (standing by functions $\sin(\omega t)$ and $\cos(\omega t)$) for each of the variables: $x_C, z_C, \varphi, x_1, x_2, z_1, z_2$, the linear system of algebraic equations with 14 unknowns had to be solved.

The aim of the calculations is to find the positions of dynamic vibration absorbers, tuned to the excitation force frequency, so that it is possible to eliminate or significantly reduce the forces transmitted to the floor by the machine supports.

As already mentioned, in practice the excitation force is mostly of the polyharmonic type. Determining the optimal positions of absorbers for each excitation frequency (in the band containing all natural frequencies of the press) will allow to find the optimal positions of absorbers designed to attenuate the selected harmonics of the excitation polyharmonic force.

The amplitudes of the horizontal reaction forces at the left and right side of the machine are always the same, while the amplitudes of the vertical reaction forces in the left and right supports do not have to be equal. Although the horizontal and vertical components of the forces in supports do not have to be in the same phase, an equivalent dimensionless reaction force is introduced in the following form:

$$\tilde{R}_{eq} = \frac{1}{|F_0|}\sqrt{R_x^2 + R_z^2} \tag{8}$$

where: $R_x = R_{xL} = R_{xR}$, $R_z = \max\{R_{zL}, R_{zR}\}$.

In the above expressions the following notations are introduced: R_{xL}, R_{xR} (R_{zL}, R_{zR}) – the amplitudes of the horizontal (vertical) components of the reaction force at the left and right side of the machine.

In the calculations the mass of the individual absorber (i.e. horizontal – m_1, vertical – m_3) equal to 300 kg was assumed, about 2.7% of the mass of the press. The stiffness of the horizontal (vertical) absorber (tuned to the excitation force frequency) is given by: $k_1 = m_1\omega^2$ ($k_3 = m_3\omega^2$).

For the case of only one absorber attached, it was assumed as the objective function the equivalent dimensionless reaction force dependent on the parameter e_1 (d_3) describing position of the horizontal (vertical) absorber: $\tilde{R}_{eq} = \tilde{R}_{eq}(e_1)$ ($\tilde{R}_{eq} = \tilde{R}_{eq}(d_3)$). The optimal positions of the absorbers were determined using the gradient method, searching for the minimum value of the objective function the following limitations were imposed: $|e_1| \le 0.8\ \text{m}$, $|d_3| \le 0.9\ \text{m}$.

The amplitudes of the dimensional forces transmitted in the horizontal and vertical directions to the supports, on the left ($R_{xL}/|F_0|$) and right side ($R_{zL}/|F_0|$, $R_{zR}/|F_0|$), for the press without the absorbers attached, are shown in Fig. 4. The diagrams in Fig. 4 present also the equivalent dimensionless reaction forces for the press without the absorbers, and for the press with the single absorber, horizontal or vertical, attached at the optimal positions.

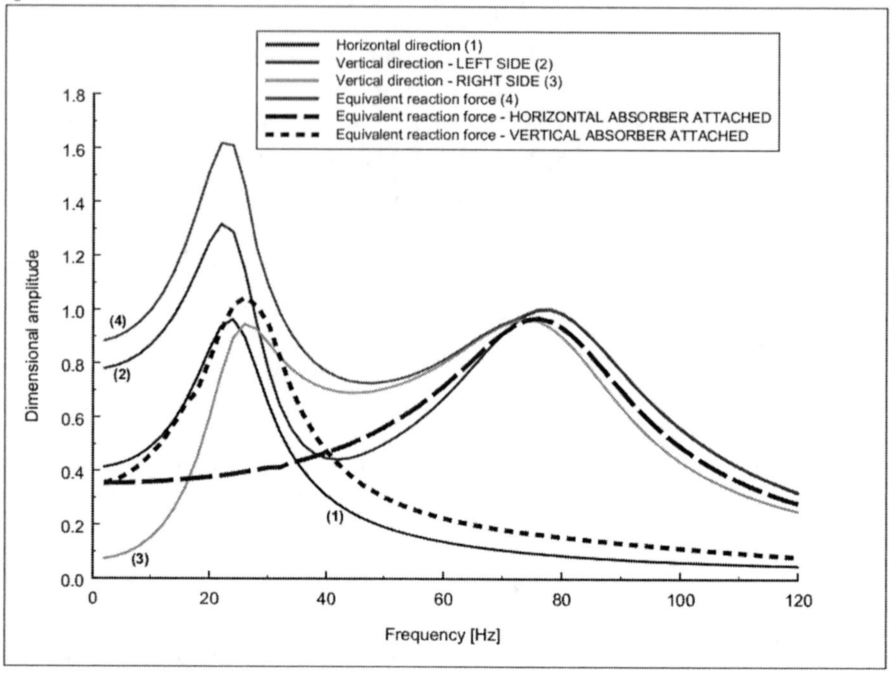

Fig. 4. Dimensionless amplitudes of the transmitted forces for the press without and with the single absorbers attached.

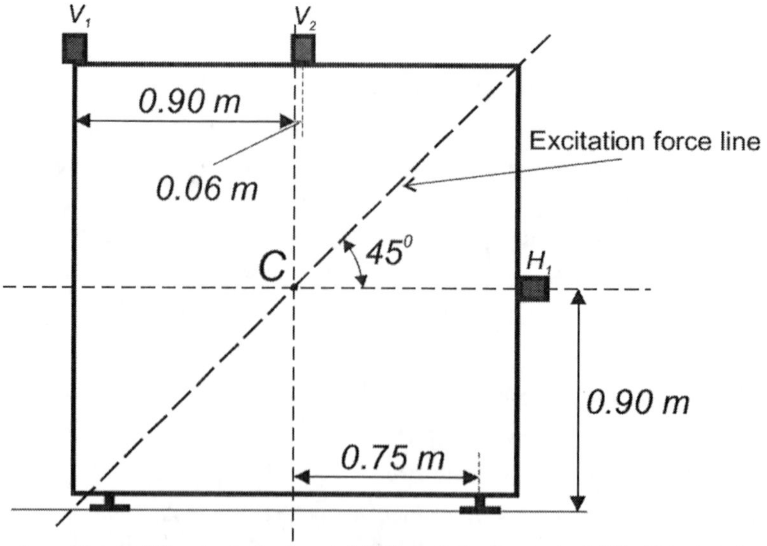

Fig. 5. Optimal positions of the single horizontal (H_1) and single vertical (V_1, V_2) absorber: H_1 – the same position for each frequency; V_1 – for: $0 < f < 30$ Hz ; V_2 – for: $f > 42$ Hz .

It is evident that the properly placed horizontal absorber is more effective up to a frequency of around 40 Hz, above this frequency the properly located vertical absorber is more effective. It turns out that the optimal position of the horizontal absorber is always the same – at the height of the machine's mass center: $e_1 = 0.0$ m (Fig. 3, Fig. 5). The optimal position of the vertical absorber depends on the frequency band: up to 30 Hz it is the extreme admissible left position: $d_3 = -0.9$ m; for the frequencies between 30 Hz and 42 Hz the optimal position lies in the range $0.04\text{ m} < d_3 < 0.06\text{ m}$; for the frequencies above 42 Hz this is the place described by the value: $d_3 = 0.07$ m (Fig. 3, Fig. 5).

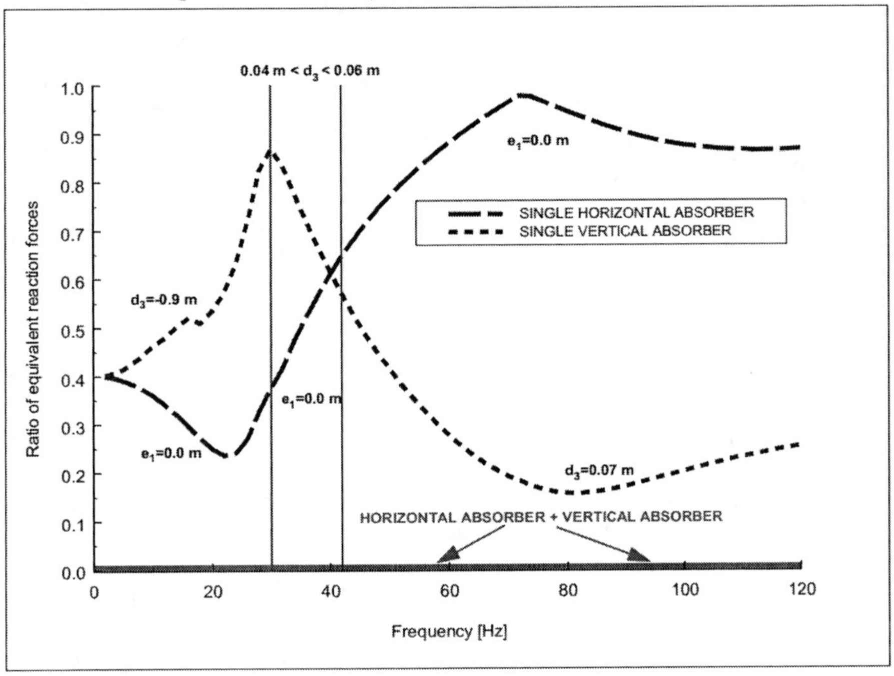

Fig. 6. Ratio of the equivalent reaction forces for the system with absorbers to its value for the machine without the absorbers attached.

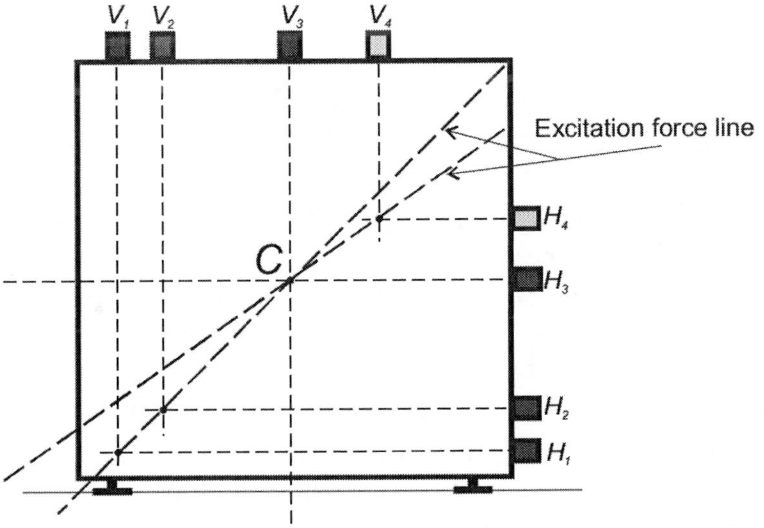

Fig. 7. Exemplary optimal positions of components of the double-absorber system $[H_i, V_i]$.

In order to better investigate effectiveness of the absorbers, in Fig. 6 the ratio of the equivalent reaction force for the system with absorbers to its value for the machine without the absorbers is shown. It is visible that the vibrations suppression efficiency of the single absorbers is limited, especially in the range between 35 Hz and 45 Hz. The highest effectiveness of the horizontal absorber is achieved around the first natural frequency of the press ($f_1 = 24\ \text{Hz}$), while for the vertical absorber around the second natural frequency ($f_2 = 78\ \text{Hz}$) which is the frequency of the vertical, uncoupled vibration mode.

The results of calculations showed that application of the system of two properly located absorbers (one horizontal and one vertical) allows to completely eliminate the vibrations of the machine. The optimal positions are such that the absorber motion lines should cross at the same point the action line of the excitation force (this ensures that the sum of moments of all forces with respect to this point is equal to zero). This conclusion is true regardless of the angle of inclination of the force line (for $0^0 < \beta_0 < 90^0$) and valid independent of the mass of the absorbers. The examples of the optimal double-absorber configurations are shown schematically in Fig. 7. For each pair of absorbers the reaction forces in the press supports are equal to zero. The choice of the best configuration may depend on additional conditions, because some positions may be excluded due to the technical (e.g. no assembly possible) or safety reasons.

5 Conclusions

In the paper the problem of the optimal distribution of dynamic vibration absorbers for the mechanical press subjected to harmonic excitation is presented. The obtained results allow to determine the optimal position of the single absorber, horizontal or vertical, and the optimal positions of the double-absorber system components, tuned to the excitation force frequency. The suppression efficiency of the single absorbers is limited, while the use of the double-absorber system, appropriately mounted, allows to completely eliminate the vibrations of the machine. The results obtained allow in practice to find the optimal positions of the absorbers tuned to the selected frequencies of harmonics of the polyharmonic excitation. It should be remembered that some location of absorbers may be limited by the technological or safety issues.

References

1. M. Abdel-Rohman, J.J. Mariam, J. Vib. Control **12(9)**, 1011–1046 (2006)
2. S.R. Chen, J. Wu, Eng. Struct. **30**, 3316–3324 (2008)
3. F. Ricciardelli, J. Wind. Eng. Ind. Aerod. **89**, 1539–1551 (2001)
4. J. Snamina, P. Martynowicz, W. Łatas, Solid State Phenomen. **208**, 29–39 (2014)
5. M. Luu, V. Zabel, C. Könke, Finite Elem. Anal. Des. **53**, 13–23 (2012)
6. Quan. Li, J. Fan, J. Nie, Quanwang. Li, Y. Chen, J. Sound Vib. **329**, 4068–4092 (2010)
7. Y. Nakano, H. Takahara, E. Kondo, J. Sound Vib. **332**, 1626–1638 (2013)
8. F. Rüdinger, Eng. Struct. **28**, 1774–1779 (2006)
9. S. Sgobba, G.C. Marano, Mech. Syst. Signal Pr. **24**, 1739–1755 (2010)
10. O.F. Tigli, J. Sound Vib. **331**, 3035–3049 (2011)
11. W. Łatas, J. Theor. App. Mech.-Pol. **53(2)**, 467–476 (2015)
12. C. Min-Li, Chi-C. Lin, Jin-M. Ueng, Kai-H. Hsieh, Jer-F. Wang, Struct. Control Health Monit. **17**, 532–548 (2010)

MATEC Web of Conferences **285**, 00009 (2019) https://doi.org/10.1051/matecconf/201928500009
DYNKON 2019

Free vibration of frame structures made of Zener type viscoelastic material

Roman Lewandowski[1,*], and *Przemysław* Wielentejczyk[2]

[1]Poznan University of Technology, Institute of Structural Engineering, 60-965 Poznan, Poland
[2]NEOKON, Construction Design Office, ul. Miodowa 12, 60-591 Poznan, Poland

Abstract. A method for determining the dynamic characteristics of structures made of viscoelastic material is presented. The fractional Zener model is used to the describe the rheological properties of materials. All of the elements of a structure must be built of material with identical rheological properties. The solution to the linear eigenvalue problem for some elastic structure and the solution to a single nonlinear algebraic equation are needed to obtain the dynamic characteristics of a viscoelastic structure. Moreover, the frequency response functions are determined in a very efficient way. The results of a representative calculation are presented and briefly discussed.

1 Introduction

The natural frequencies, non-dimensional damping ratios, modes of vibration and frequency response functions are very important characteristics of each mechanical structure. For elastic systems, these properties can be determined using well-known methods, described in many monographs (see, for example [1, 2]). However, methods which are devoted to determining the dynamic properties of viscoelastic (VE) structures or elastic structures with viscoelastic dampers also exist (see, for example [3 - 5]) but they do not fully clarify a number of problems. For this type of structures, the dynamic characteristics are obtained by solving a nonlinear eigenvalue problem. In some cases, when the VE materials or dampers are described using the classic rheological models, the above-mentioned nonlinear eigenvalue problem can be transformed into the linear one after introducing the concept of state and/or internal variables [2, 6]. The methods used to solve the nonlinear eigenvalue problem include the continuation method [3, 4], Newton's eigenvalue iteration method [7], the iterative method [5] and some other ones.

In this paper, the dynamic characteristics of VE frame structures made of a material described by the fractional Zener model are analysed. It is assumed that the whole structure is homogeneous and made of a material described by a fractional Zener model and the parameters of such model are identical for all the elements of the structure. In particular, the new method for solving the nonlinear eigenvalue problem, of which the computational efficiency is very high, is presented. In the method, it is necessary to know a solution to the linear eigenvalue problem for some elastic structures and a solution to a single algebraic equation to determine the dynamic characteristics of the VE structures. However, the

* Corresponding author: roman.lewandowski@put.poznan.pl

presented method can be used only to analyse the above-mentioned homogeneous VE structures. The frequency response functions (FRF) are determined in a very efficient way.

2 Equation of motion for a rod treated as a continuous system

The Euler theory is used to describe the dynamic behaviour of rods made of viscoelastic (VE) materials. The well-known equilibrium equations are:

$$-N_{,x}(x,t)+m\ddot{u}(x,t)=p_x(x,t) \ , \tag{1}$$

$$-T_{,x}(x,t)+m\ddot{w}(x,t)=p_y(x,t) \ , \qquad T(x,t)=M_{,x}(x,t) \ , \tag{2}$$

where $M(x,t)$, $T(x,t)$ and $N(x,t)$ are the bending moments, the shear and normal forces, respectively; $u(x,t)$, $w(x,t)$ are the axial and transversal displacements of the neutral axis of the rods; $p_x(x,t)$ and $p_y(x,t)$ are the axial and transversal excitation forces per unit length of a rod and m is the mass per unit length of the rod. The dot indicates differentiation with respect to time t and $(\bullet)_{,x}(x,t)=d(\bullet)/dx$.

The axial strain of the neutral axis of the rod $\varepsilon(x,t)$, the angle of rotation of the rod's cross-section $\varphi(x,t)$ and the rod's curvature $\kappa(x,t)$ are defined as:

$$\varphi(x,t)=w_{,x}(x,t) \ , \qquad \kappa(x,t)=-w_{,xx}(x,t) \ , \qquad \varepsilon(x,t)=u_{,x}(x,t) . \tag{3}$$

The constitutive relationships for the viscoelastic material of Zener type written at the rod's cross-section level take the following form [8]:

$$M(x,t)+\tau^{\alpha}\mathrm{D}_t^{\alpha}M(x,t)=E_0J\kappa_x(x,t)+\tau^{\alpha}E_{\infty}J\mathrm{D}_t^{\alpha}\varepsilon_x(x,t) \ , \tag{4}$$

$$N(x,t)+\tau^{\alpha}\mathrm{D}_t^{\alpha}N(x,t)=E_0A\varepsilon(x,t)+\tau^{\alpha}E_{\infty}A\mathrm{D}_t^{\alpha}\varepsilon(x,t) \ , \tag{5}$$

where A and J are the rod's cross-sectional area and the moment of inertia of the cross-section. Moreover, E_0, E_{∞} and τ are the relaxed and non-relaxed elastic modulus and the relaxation time, respectively. $\mathrm{D}_t^{\alpha}x(t)$ is the Riemann-Liouville fractional derivative of order α of function $x(t)$ with respect to time.

The governing relationships of the non-viscously damped systems are often defined in the frequency domain. For this reason, Eqns (4) and (5) can be equivalently expressed in the Laplace domain (assuming zero initial conditions) as follows:

$$\overline{N}(x,s)=s\overline{G}(s)A\overline{\varepsilon}(x,s) \ , \qquad \overline{M}(x,s)=s\overline{G}(s)J\overline{\kappa}(x,s) \ , \tag{6}$$

where s is the Laplace variable and $s\overline{G}(s)$ is the Laplace transform of the heredity function. The particular form of the $s\overline{G}(s)$ function for the Zener model of material is:

$$s\overline{G}(s)=\left(1+\frac{\tau^{\alpha}s^{\alpha}}{1+\tau^{\alpha}s^{\alpha}}\frac{\Delta E}{E_0}\right)E_0=(1+\theta(s))E_0 \ , \tag{7}$$

where $\Delta E=E_{\infty}-E_0$.

The Laplace transformation of Eqs (1) – (3) give us:

$$-\overline{N}_{,x}(x)+s^2m\overline{u}(x)=\overline{p}_x(x)\ ,\qquad -\overline{T}_{,x}(x)+s^2m\overline{w}(x)=\overline{p}_y(x)\ ,\tag{8}$$

$$\overline{T}(x)=\overline{M}_{,x}(x)\ ,\tag{9}$$

$$\overline{\varphi}(x)=\overline{w}_{,x}(x)\ ,\qquad \overline{\kappa}(x)=-\overline{w}_{,xx}(x)\ ,\qquad \overline{\varepsilon}(x)=\overline{u}_{,x}(x)\ ,\tag{10}$$

where the quantities with the bar are the Laplace transforms of the corresponding quantities without the bar.

3 The finite element formulation for a viscoelastic rod

The finite element method is a general and widely-accepted method to solve many dynamic problems. Here, the planar, two-node rod element with six nodal parameters and the Hermite shape functions, shown in Fig. 1, are used to describe the dynamic behaviour of a viscoelastic rod.

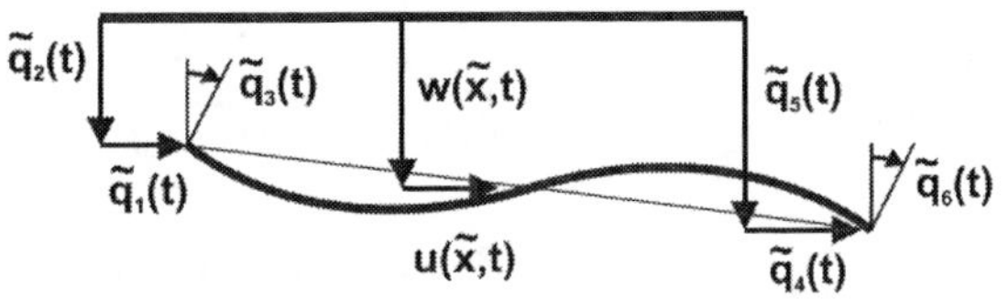

Fig. 1. Schematic of the planar finite element.

The vector of nodal parameters (in a local coordinate axis) is $\mathbf{q}_e(t)=[\tilde{q}_1(t),\ \tilde{q}_2(t),\ \tilde{q}_3(t),\ \tilde{q}_4(t),\ \tilde{q}_5(t),\ \tilde{q}_6(t)]^T$ (see Fig.1). The vector of displacements defined in the time domain is $w_e(x,t)=[u(x,t),\ w(x,t)]^T$ and defined in the Laplace domain $\overline{\mathbf{w}}_e(x,s)=[\overline{u}(x,s),\overline{w}(x,s)]^T$ is written in terms of nodal parameters in the following way:

$$\mathbf{w}_e(x,t)=\mathbf{H}(x)\mathbf{q}_e(t)\ ,\qquad \overline{\mathbf{w}}_e(x,s)=\mathbf{H}(x)\overline{\mathbf{q}}_e(s)\ ,\tag{11}$$

where $\mathbf{H}(x)$ is the well-known matrix of shape functions [2]. The vector of generalized strains $\mathbf{e}(x,t)=[\varepsilon(x,t),\kappa(x,t)]^T$ and his Laplace transform $\overline{\mathbf{e}}(x,s)=[\overline{\varepsilon}(x,s),\overline{\kappa}(x,s)]^T$ depends on the nodal parameters as follows:

$$\mathbf{e}(x,t)=\mathbf{B}(x)\mathbf{q}_e(t)\ ,\qquad \overline{\mathbf{e}}(x,s)=\mathbf{B}(x)\overline{\mathbf{q}}_e(s)\ ,\tag{12}$$

where $\mathbf{B}(x)$ is the matrix containing appropriate derivatives of the shape functions [2].

In both of the above-mentioned domains, the vector of excitation forces is defined as $\mathbf{p}_e(x,t)=[p_x(x,t),p_y(x,t)]^T$ and $\overline{\mathbf{p}}_e(x,s)=[\overline{p}_x(x,s),\overline{p}_y(x,s)]^T$. Moreover, the vectors of inertia forces $\mathbf{b}_e(x,t)=[b_x(x,t),b_y(x,t)]^T$ and $\overline{\mathbf{b}}_e(x,s)=[\overline{b}_x(x,s),\overline{b}_y(x,s)]^T$ depend on nodal acceleration in the usual way

$$\mathbf{b}_e(x,t)=-m\mathbf{H}(x)\ddot{\mathbf{q}}_e(t)\ ,\qquad \overline{\mathbf{b}}_e(x,s)=-s^2m\mathbf{H}(x)\overline{\mathbf{q}}_e(s)\ .\tag{13}$$

After introducing the vector of generalized stresses $\mathbf{s}(x,t)=[N(x,t),M(x,t)]^T$, the matrix form of the constitutive relationships (6) is:

$$\overline{\mathbf{s}}(x,s)=s\overline{\mathbf{G}}(s)\overline{\mathbf{e}}(x,s)\ ,\tag{14}$$

where

$$s\overline{\mathbf{G}}(s) = s\overline{G}(s)\begin{bmatrix} A & 0 \\ 0 & J \end{bmatrix} . \tag{15}$$

Using the above notations, the virtual work equation for the finite element could be written as follows:

$$\delta\mathbf{q}_e^T(t)\mathbf{R}_e(t) + \int_0^{l_e} \delta\mathbf{w}_e^T(x,t)(\mathbf{p}_e(x,t) + \mathbf{b}_e(x,t))dx = \int_0^{l_e} \delta\mathbf{e}^T(x,t)\mathbf{s}(x,t)dxdt , \tag{16}$$

where the symbols $\delta\mathbf{w}_e(x,t)$, $\delta\mathbf{q}_e(t)$ and $\delta\mathbf{e}(x,t)$ denote the vector of virtual displacements, the vector of virtual nodal displacements and the vector of virtual generalized strains, respectively. Moreover, $\mathbf{R}_e(t) = [R_1(t), R_2(t), R_3(t), R_4(t), R_5(t), R_6(t)]^T$ is the vector of nodal reactions and l_e is the length of the finite element.

In the Laplace's domain, the virtual work equation (16) takes the following form:

$$\delta\overline{\mathbf{q}}_e^T\overline{\mathbf{R}}_e + \int_0^{l_e} \delta\overline{\mathbf{w}}_e^T(x)(\overline{\mathbf{p}}_e(x) + \overline{\mathbf{b}}_e(x))dx = \int_0^{l_e} \delta\overline{\mathbf{e}}^T(x)\overline{\mathbf{s}}(x)dx . \tag{17}$$

Proceeding in the usual way, the second and third terms of Eq. (17) can be rearranged in the following way:

$$\int_0^{l_e} \delta\overline{\mathbf{w}}_e^T(x)\overline{\mathbf{p}}_e(x)dx = \delta\overline{\mathbf{q}}_e^T\overline{\mathbf{P}}_e , \qquad \int_0^{l_e} \delta\overline{\mathbf{w}}_e^T(x)\overline{\mathbf{b}}_e(x)dx = -s^2\delta\overline{\mathbf{q}}_e^T\mathbf{M}_e\overline{\mathbf{q}}_e , \tag{18}$$

where $\overline{\mathbf{P}}_e$ is the Laplace transform of nodal excitation forces and $\mathbf{M}_e$ is the well-known elemental mass matrix of the rod. Moreover, the right hand size of Eq. (17) could be written, using relationships (12.2) and (14), in the form:

$$\int_0^{l_e} \delta\overline{\mathbf{e}}^T(x)\overline{\mathbf{s}}(x)dx = \delta\overline{\mathbf{q}}_e^T\widetilde{\mathbf{K}}_e\overline{\mathbf{q}}_e , \tag{19}$$

where the elemental viscoelastic stiffness matrix is

$$\widetilde{\mathbf{K}}_e = \int_0^{l_e} \mathbf{B}^T(x)s\overline{\mathbf{G}}(s)\mathbf{B}(x)dx . \tag{20}$$

It is useful to divide the above viscoelastic stiffness matrix into two matrixes: the elemental elastic stiffness matrix $\mathbf{K}_e$ and the second matrix $\mathbf{K}_{ev}$ which will be called the elemental viscous matrix. To be exact, the last matrix expresses the damping properties of the rod only when the classic Zener model (i.e. when $\alpha = 1$ in Eq. (4) and (5)) is used in the constitutive equation (14). Taking into account relationships (7), it is possible to write

$$\widetilde{\mathbf{K}}_e = \mathbf{K}_e + \mathbf{K}_{ev} = (1 + \theta(s))\,\mathbf{K}_e , \tag{21}$$

where

$$\mathbf{K}_{ev} = \frac{\tau^\alpha s^\alpha}{1+\tau^\alpha s^\alpha} \frac{\Delta E}{E_0} \int_0^{l_e} \mathbf{B}^T(x)\mathbf{E}_0\mathbf{B}(x)dx \ , \qquad \mathbf{K}_e = \int_0^{l_e} \mathbf{B}^T(x)\mathbf{E}_0\mathbf{B}(x)dx \ , \tag{22}$$

are the above-mentioned elemental viscous matrix and the well-known elemental stiffness matrix of the elastic rod made of an elastic material of which the elastic module is E_0, respectively. Moreover, $\mathbf{E}_0 = E_0 diag(A,\ J)$. It is important to note that both the above-mentioned matrices are proportional and the factor of proportionality is $\theta(s)$.

Finally, from the virtual work equation (17), the following equation for the finite element

$$\overline{\mathbf{R}}_e = (1+\theta(s))\mathbf{K}_e\overline{\mathbf{q}}_e + s^2\mathbf{M}_e\overline{\mathbf{q}}_e - \overline{\mathbf{P}}_e \ , \tag{23}$$

is obtained.

4 Dynamic properties of viscoelastic frame structures

Now it is assumed that the whole frame structure is made of a homogenous viscoelastic material and one rheological model is used to describe the mechanical properties of the material. Applying the usual finite element procedure, the equation of motion of structure in the Laplace domain can be written in the following form:

$$\left\{[1+\theta(s)]\mathbf{K} + s^2\mathbf{M}\right\}\overline{\mathbf{q}} = \mathbf{P} \ , \tag{24}$$

where $\overline{\mathbf{q}}$ and $\overline{\mathbf{P}}$ are global vectors of the Laplace transform of nodal displacements and excitation forces, respectively. Moreover, $\mathbf{M}$ and $\mathbf{K}$ are the global mass matrix and the global stiffness matrix of the elastic frame, respectively. The stiffness matrix $\mathbf{K}$ is built using the elemental matrices given by Eq. (22.2).

The dynamic properties of viscoelastic structure are obtained assuming, firstly, that $\overline{\mathbf{P}} = \mathbf{0}$, and solving the nonlinear eigenvalue problem

$$\left\{[1+\theta(s)]\mathbf{K} + s^2\mathbf{M}\right\}\overline{\mathbf{q}} = \mathbf{0}, \tag{25}$$

where s is the eigenvalue and $\overline{\mathbf{q}}$ is the eigenvector.

In a general case and when the properties of viscoelastic elements of structure are modelled with the help of classic rheological models, the modes of vibration are complex and are also called the viscous or elastic modes of vibration, except that, for some rheological models (for example: the Zener and the generalized Maxwell model), the real mode of vibration, also called the non-viscous mode, exists. However, for the fractional Zener model, the real mode of vibration does not exist [3].

It is difficult to solve the nonlinear eigenvalue problem of this type. Fortunately, in the considered case the eigenvalue problem (25) can easily be transformed into the following linear eigenvalue problem

$$(\mathbf{K} - \omega_e^2\mathbf{M})\,\overline{\mathbf{q}}_l = \mathbf{0} \ , \tag{26}$$

where $\overline{\mathbf{q}}_l$ is the eigenvector and ω_e the eigenvalue of the linear eigenvalue problem (26) and linked with the eigenvalue of the nonlinear problem by

$$f(s) = s^2 + \omega_e^2(1+\theta(s)) = 0 \ . \tag{27}$$

It is easy to recognize that the linear eigenvalue problem (26) is identical to the eigenvalue problem from which the natural frequency of the considered structure is calculated but is made of an elastic material with the elastic modulus E_0. Many well-known procedures [1] can be used to solve the linear eigenvalue problem (26).

The matrices $\mathbf{K}$ and $\mathbf{M}$ in Eqns (25) and (26) are positive definite ones, which means that the eigenvalue ω_e^2 is a real, positive number and the eigenvector $\overline{\mathbf{q}}_l$ is a real vector. Moreover, since $\overline{\mathbf{q}} = \overline{\mathbf{q}}_l$, the eigenvector of the nonlinear eigenvalue problem (25) is also a real vector. In addition, two eigenvectors of this kind (say, $\overline{\mathbf{q}}_i$ and $\overline{\mathbf{q}}_j$), fulfil the well-known orthogonality conditions:

$$\overline{\mathbf{q}}_i^T \mathbf{M}\, \overline{\mathbf{q}}_j = \delta_{ij} \,, \qquad \overline{\mathbf{q}}_i^T \mathbf{K}\, \overline{\mathbf{q}}_j = \omega_{ei}^2 \delta_{ij} \,, \tag{28}$$

where δ_{ij} is the Kronecker's symbol.

However, the eigenvalues of the above mentioned problems (25) and (26) are different because s must be determined from Equation (27). It is a nonlinear algebraic equation and, for a given ω_e, it has more than one solution. Numerical experiments show that Eq. (27) has two complex, conjugate solutions in the case $\alpha \neq 1$ and three solutions (two complex, conjugate solutions and one real solution) when $\alpha = 1$.

In the case of elastic structures with n-th degree of freedom, we have n eigenvalues and eigenvectors but there are more than n solutions to the eigenvalue problem for viscoelastic structures. When the fractional Zener ($\alpha \neq 1$) is used, it is $2n$ eigenvalues but in the case of the classic Zener model ($\alpha = 1$) we have $3n$ eigenvalues. Because the number of eigenvectors $\overline{\mathbf{q}}$ and $\overline{\mathbf{q}}_l$ is n, we conclude the vector $\overline{\mathbf{q}}$ is in fact a repeated eigenvector corresponding to two or three eigenvalues s determined from Eq. (27) for a given value of ω_e.

Having a complex solution $s = \mu + \mathrm{i}\eta$ to Eq. (25), it is possible to calculate the natural frequency ω and the non-dimensional damping ratio γ of the viscoelastic structure using the following formulae:

$$\omega^2 = \mu^2 + \eta^2 \,, \qquad \gamma = -\mu / \omega \,. \tag{29}$$

Often the frequency response functions (FRF) are used to characterize the dynamic behaviour of structures. The matrix of FRF is obtained assuming $s = \mathrm{i}\lambda$ and inverting the matrix $\mathbf{D}(\lambda) = [1 + \theta(\mathrm{i}\lambda)]\mathbf{K} - \lambda^2 \mathbf{M}$. For the considered type of structures, this can be done either in a direct way or using a concept similar to that of normal coordinates as described below. For large systems, the second approach is more efficient.

The solution to Eq. (24) for $s = \mathrm{i}\lambda$ is assumed in the form:

$$\overline{\mathbf{q}}(\mathrm{i}\lambda) = \sum_{j=1}^{n} \mathbf{a}_j \overline{x}_j(\lambda) \,, \tag{30}$$

where $\mathbf{a}_j$ is the eigenvector fulfilling Eq. (26) and $\overline{x}_j(\lambda)$ is the Fourier transform of normal coordinate $x_j(t)$ of the elastic structure. After introducing relationship (30) into Eq. (24) multiplied by $\mathbf{a}_i^T$ and taking into account the orthogonality conditions (28), the following equation is obtained:

$$\left\{[1 + \theta(\mathrm{i}\lambda)]\omega_{ei}^2 - \lambda^2\right\}\overline{x}_i = \mathbf{a}_i^T \mathbf{P} \,, \tag{31}$$

from which

$$\bar{x}_i = \frac{\mathbf{a}_i^T \mathbf{P}}{[1+\theta(\mathrm{i}\lambda)]\omega_{ei}^2 - \lambda^2} \, . \tag{32}$$

The matrix of FRF $\mathbf{Z}(\mathrm{i}\lambda) = \left\{[1+\theta(\mathrm{i}\lambda)]\mathbf{K} - \lambda^2 \mathbf{M}\right\}^{-1}$ can now be rewritten in the form:

$$\mathbf{Z}(\mathrm{i}\lambda) = \sum_{j=1}^{n} \frac{\mathbf{a}_j \mathbf{a}_j^T}{[1+\theta(\mathrm{i}\lambda)]\omega_{ej}^2 - \lambda^2} \, . \tag{33}$$

5 Results of a representative calculation

The two span beam, shown in Fig. 2, is considered an example. The following data are adopted: the span of the beam $L = 4.0\ \mathrm{m}$, the dimension of the beam's cross-section $b = 0.2\ \mathrm{m}$, $h = 0.4\ \mathrm{m}$, the unit mass $m = 80.0\ \mathrm{kg/m}$, the parameters of the Zener model are $E_0 = 7.0\ \mathrm{MPa}$, $\tau = 0.02\ \mathrm{s}$ and $\alpha = 0.7$. The calculation is performed for different ratios E_∞ / E_0 taken in a wide range of $1.1 \div 10.0$. The elastic beam corresponding to the VE one has the modulus $E = 7.0\ \mathrm{MPa}$.

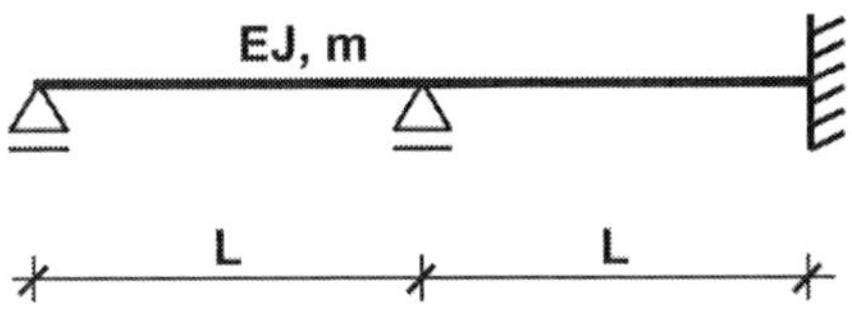

Fig. 2 Diagram of the two-span beam.

The first two natural frequencies of the elastic beam are: $\omega_1 = 6.85\ \mathrm{rad/s}$ and $\omega_2 = 11.87\ \mathrm{rad/s}$. The Newton method is used to solve the nonlinear equation (27). Calculations are performed for different values of E_∞ and the results of the calculations are presented in Figs 3 and 4. In Fig. 3, the non-dimensional damping ratio of the first mode (solid line) and of the second mode (dashed line) is presented as a function of the modulus ratio E_∞ / E_0. Both ratios grow first with the E_∞ / E_0 ratio but the non-dimensional damping ratio is approximately constant for ratios higher than 12.0 (for the first mode) and 10.0 (for the second mode).

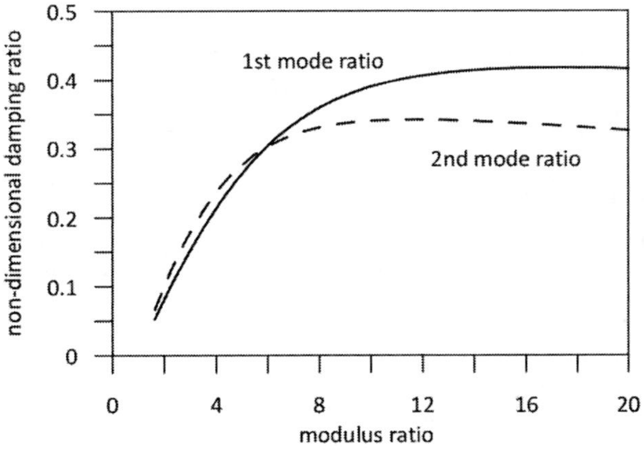

Fig. 3 The non-dimensional damping ratios versus the modulus ratio E_∞ / E_0.

In Fig. 4, changes of the first and second natural frequencies versus modulus ratio E_∞ / E_0 are shown. Both natural frequencies grow significantly with increasing modulus ratios.

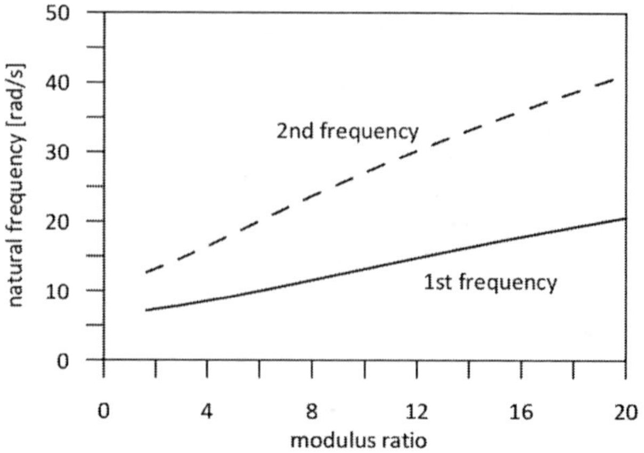

Fig. 4 The natural frequencies versus the modulus ratio E_∞ / E_0.

Finally, it must be stressed that the VE material of the considered beam has very good damping properties.

6 Concluding remarks

A method for determining the dynamic characteristics of homogenous VE frame structures, which is computationally very efficient, is presented. The method requires a solution to the linear eigenvalue problem for corresponding elastic structures and a solution to one nonlinear algebraic equation. This makes the method very attractive from the computational point of view. The method is not general but can be applied to a technically important class of structures. It is obvious that the method could also be applied also to other rheological models of VE materials.

This study has been supported by the Poznan University of Technology, the Institute of Structural Engineering, carried out in the year 2019. This support is gratefully acknowledged.

References

1. K.J. Bathe, *Finite Element Procedures in Engineering Analysis*, (Prentice-Hall, 1982)
2. R. Lewandowski, *Dynamika konstrukcji budowlanych*, 2 Ed., (Wydawnictwo Politechniki Poznańskiej, 2018)
3. R. Lewandowski, M. Baum, Arch. Appl. Mech., **85**, 1793–1814, (2015)
4. Z. Pawlak, R. Lewandowski, Comput. Struct. **125**, 53–61, (2013)
5. S. Adhikari, B. Pascual, J. Vib. Acoust., **132,** 021002.1–021002.7. (2011)
6. R. Lewandowski, A. Bartkowiak, H. Maciejewski, Struct. Eng. Mech., **41**, 113–137 (2012)
7. K.V. Singh, Int. J. Mech. Sci., **110**, 127–137, (2016)
8. R. Lewandowski, P. Wielentejczyk, J. Sound Vibr., **399**, 228–243 (2017)
9. F. Mainardi and G. Spada, Creep, relaxation and viscosity properties for basic fractional models in rheology, Eur. Phys. J. Special Topics 193, 133–160 (2011)

Experimental verification of the numerical model of a reinforced concrete arch

Barbara Markiewicz[1,*], *Grzegorz* Piątkowski[1] , *Łukasz* Szyszka[1], and *Dominika* Ziaja[1]

[1]Rzeszow University of Technology, The Faculty of Civil and Environmental Engineering and Architecture, Poznanska 2, 35-084 Rzeszow, Poland

Abstract. The paper presents numerical and experimental dynamic analysis of the reinforced concrete arch. The measurement of the accelerations at various points of the structure as a response to the force (single input-multiple output approach) was made. Dynamic response of the concrete arch has been measured by piezoelectric accelerometers and signals have been recorded by multi-channel LMS SCADAS Mobile recorder. As a result of the measurements frequencies and mode shapes have been determined. ADINA program has been used for building of numerical models. The static scheme of the arch (the stiffness of the supports) has been updated using the genetic algorithms (GA) available in MATLAB to obtain the results, which correspond to dynamic testing. Two similarity criteria have been considered: the mean squared error (MSE) between frequencies and the Modal Assurance Criterion (MAC) for forms comparison. In additional example, the Young's modulus has been also optimized.

1 Introduction

Nowadays we can clearly observe the increase of the importance of dynamic analysis in structural design. The sensitivity to dynamic loads, i.e. wind pressure, earthquake, machine operation, road and rail traffic, is of great importance. These loads, even at small values, can endanger the safety and durability of a structure, especially in the case of slender and light constructions. Dynamic analysis consider not only the stiffness, but also the mass of the structure, thus giving a more accurate information of the behaviour of the structure.

What is more, dynamic analysis can be successfully used in structural assessment procedures. Existence of a structural damage leads to modification of modal parameters (natural frequencies, mode shapes and modal damping values). Changes in the modal parameters may not be the same for each mode. They depend on the nature, location and severity of the damage. Thus, an analysis of periodical data from dynamic testing can be used in monitoring of structural condition [1].

Going further, structural monitoring and testing procedures for the dynamic characterization of a structure are essential tools for the calibration of numerical models. The process of matching parameters of numerical model to the target response, so called

* Corresponding author: dziaja@prz.edu.pl

model updating, implements the optimization to identify parameters that minimize function representing the distance between the measured data and the predicted data [2].

The most popular method is a gradient-based optimization strategy. For more specific results response surface modelling (RSM), particle swarm optimization (PSO), Monte Carlo optimization, and genetic algorithms (GA) can be used. Recently, various type of the Artificial Neural Networks (ANNs) have been increasingly used as a model updating tool, especially in case of updating of computational models on the basis of limited number of measurement data [3].

In this paper numerical and experimental dynamic analysis of the reinforced concrete arch were presented. The dynamic response of the concrete arch to the impact force has been measured by accelerometers and as a result of the measurements frequencies and mode shapes have been determined. The aim of the study was to find the static scheme of numerical model, for which the experimental results could be considered as corresponding. The static scheme of the arch (the stiffness of the supports) and additionally the Young's modulus have been optimized using the genetic algorithms (GA) available in MATLAB environment. The considered similarity criteria included frequencies and modal shapes (MAC) comparison.

2 Description of the tested structure

The tested structure is a reinforced concrete arch of a span of about 10m and rise above 2m. The exact dimensions of the arch and its cross-section are shown in Figure 1.

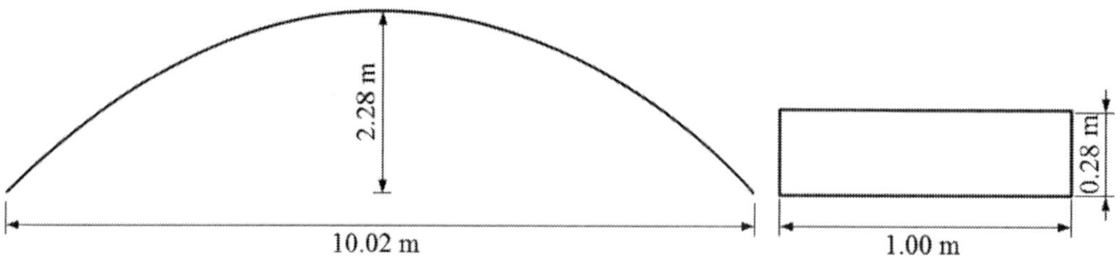

Fig. 1. Dimensions of the arch and its cross-section.

The arch is essentially a curved beam, which great advantage is the possibility of all sections of the arch to be under simple compression with no tension. This is especially important in case of reinforced concrete construction.

Arches are used for centuries for structures, where long spans are desired, in roofs of such buildings as hangars, auditoriums, gymnasiums and to support bridge decks. However, it can be noticed that they are also one of the key components for modern complex structures [4].

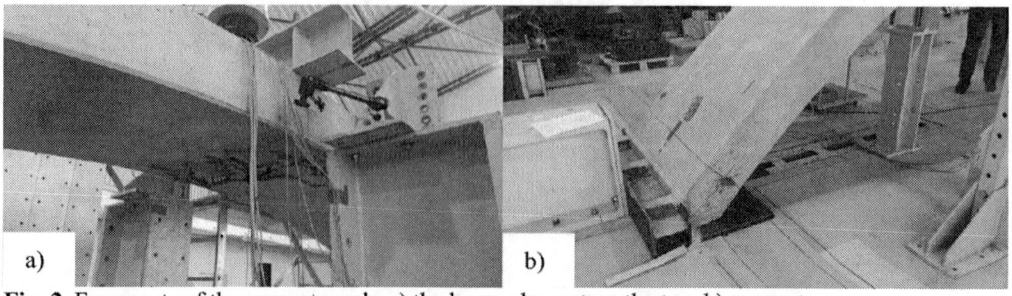

Fig. 2. Fragments of the concrete arch: a) the beam element on the top, b) support.

3 Experimental measurements

The measurements shown in this paper are only addition for bigger project realised in Rzeszow University of Technology. The main aim of the project is not an object of this paper, so the description of it has been omitted.

The classical modal analysis has been used, which is based on the measurement of the existing force and accelerations at various points of the structure as a response to this force (SIMO). Impact excitation has been generated by PCB Piezotronics modal hammer in point shown in Fig. 3, in the vertical direction.

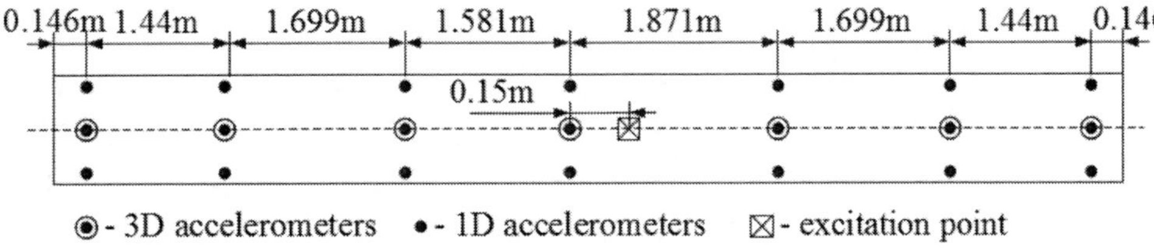

Fig. 3. Measurement points location.

Dynamic response of the concrete arch has been measured by 21 piezoelectric accelerometer sensors (two types of sensors were used: 3D and 1D). The 3D measurement points (7 sensors) have been located on the upper surface of the arch on its axis. The 1D measurement points have been placed 2cm from edges on the both side of the structure (14 sensors) to obtain a torsional form optionally. In this study, only 3D sensors data was taken into account, because only the forms in the plane were analysed. Multi-channel signal recorder Scadas Mobile's LMS International has been used in measurements. Signals in the frequency range 0-256 Hz with a resolution of 0.125 Hz have been recorded during acceleration measurements. Recorder software allows for the classical experimental modal analysis including estimation of poles using Polymax method and visualization of the results. As a result of the measurements frequencies and vibration modes have been determined.

4 Model updating

4.1 Numerical models

The ADINA program has been used for building of numerical models. Preliminary estimates of the dynamic properties are based on the simplest numerical model. The model of a studied arch is a frame element with three degrees of freedom to analyse only the in plane shape modes.

Because the method of fixing of the arch was ambiguous, see Fig. 2 b), four types of boundary conditions have been considered: clamped (NUM1), pinned at both ends (NUM2), simple support (NUM3) and clamped with one unfixed horizontal translation (NUM4), see Fig. 4. These four support variants were analysed to determine the range of expected values of the frequency and to observe dynamic behaviour of the arch depending on enabling the support to slide and/or rotate. The cases with additional 'M' include concentrated mass located on the top of the arch. This is the mass of the traverse beam made of the HEA240 profile, of 1.5m length, which is 90 kg, see Fig. 2 a).

The data used in calculations: the Young's modulus equal to 36 GPa, shear modulus 15 GPa and specific weight 24.53 kN/m^3.

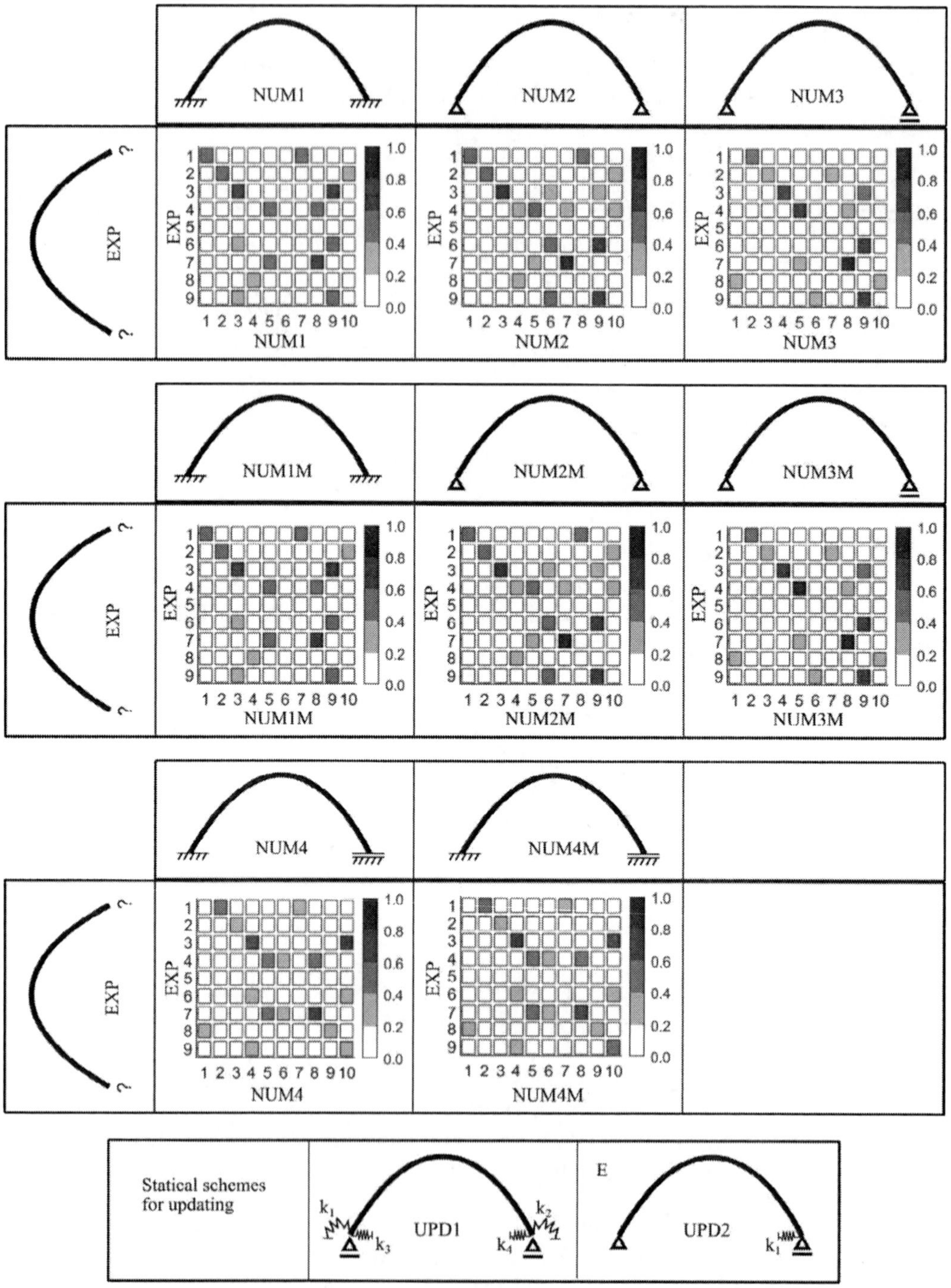

Fig. 4. The analysed schemes and correlation matrices between the analysed numerical and experimental forms

4.2 Similarity criteria

The eigenfrequencies and eigenforms of numerical models have been analysed and compared with modal frequencies and modal forms of a physical model.

Table 1. Frequencies of analysed models [Hz]

EXP	NUM1	NUM1 M	NUM2	NUM2 M	NUM3	NUM3 M	NUM4	NUM4 M
16.32	20.82	20.77	12.43	12.40	3.33	3.30	7.32	7.25
32.16	38.51	37.68	28.80	28.28	13.82	13.77	22.05	21.98
66.81	72.02	72.02	56.21	56.21	33.49	33.13	45.59	45.06
103.32	80.73	80.72	77.58	77.52	57.95	57.94	72.96	72.96
112.73	115.74	114.11	103.77	102.24	92.54	91.69	110.35	109.85
145.94	149.34	149.03	127.03	126.87	115.05	114.26	118.14	116.74
189.72	190.11	188.86	180.30	178.85	143.33	143.19	161.78	161.77
230.82	206.57	204.74	189.77	188.64	180.54	179.07	205.53	203.83
246.57	266.95	266.79	241.27	240.83	238.89	238.75	257.20	255.08
-	322.74	319.37	290.43	287.18	257.29	254.98	269.19	268.97

In Table 1 the obtained following frequencies of analysed models have been listed. In column ‘EXP’ the experimental frequencies are listed and the remaining columns are numerical frequencies for various support schemes described before. The similarity of the forms have not been considered in Table 1. As a similarity criterion of the frequencies the percentage difference has been used, calculated as in formula:

$$\Delta f_{ine} = \frac{f_{in} - f_{ie}}{f_{ie}} \cdot 100\,\% \quad , \tag{1}$$

where: Δf_{ine} - the percentage difference between numerical and experimental the i-th frequency values, f_{in} – the i-th frequency obtained in numerical simulations, f_{ie} – the i-th frequency measured during dynamic testing.

And to asses compatibility of the numerical models and experimental model based only on the frequencies values the mean squared error (MSE) has been used, according to [5] computed as:

$$err = \sum_{m} (f_{in} - f_{ie})^2 \quad , \tag{2}$$

where: err - the mean squared error, m- number of identified frequencies (9 in this case).

The similarity of obtained mode shapes has been not only visually assessed, but also parametrically using the Modal Assurance Criterion (MAC). The correlation between the vectors describing modal shapes obtained in test and numerical eigenvectors for FEM models is accomplished by means of the MAC value [6]:

$$MAC(i,j) = \frac{(\Psi_i^{nT} \Psi_j^{e})^2}{(\Psi_i^{nT} \Psi_i^{n}) \cdot (\Psi_j^{eT} \Psi_j^{e})} \tag{3}$$

where: index i - number of mode of numerical model (n), index j – number of mode of physical model (e).

According to information in [5] and [6]: 0 < MAC <1, 1 means full fit, 0 means no correlation between the mode shapes. In practice, full adjustment of numerical and experimental forms turns out to be impossible and MAC > 0.8 is considered to be in good agreement.

Figure 4 shows the correlation matrices for the analysed numerical and experimental forms. The numbers represent successive forms corresponding to the frequencies listed in Table 1. The value of the MAC coefficient between each of the two forms is symbolically represented by a square located in column and row marked with the numbers of the compared mode shapes, in the appropriate colour, according to the scale shown on the right.

Table 2 shows the matching frequencies in term of the similarity of mode shapes, which the MAC factor exceeded the value of 0.5. The similarity criterion was lowered, because it was not possible to obtain satisfactory similarity of all forms (MAC > 0.8). In Table 2 and Table 3 'N' means number of the mode.

Table 2. Comparison of experimental and numerical frequencies in term of mode shape [Hz]

EXP	**NUM1**			**NUM1M**		**NUM2**			**NUM2M**		
f [Hz]	**N [-]**	**f [Hz]**	**MAC [-]**	**f [Hz]**	**MAC [-]**	**N [-]**	**f [Hz]**	**MAC [-]**	**N [-]**	**f [Hz]**	**MAC [-]**
16.32	1	20.82	0.562	20.77	0.561	1	12.43	0.501	1	12.40	0.501
32.16	2	38.51	0.585	37.68	0.586	2	28.80	0.576	2	28.28	0.576
66.81	9	266.95	0.658	266.79	0.662	3	56.21	0.679	3	56.21	0.679
103.32	5	115.74	0.595	114.11	0.594	-	-	-	-	-	-
112.73	-	-	-	-	-	-	-	-	-	-	-
145.94	-	-	-	-	-	9	241.27	0.658	9	240.83	0.659
189.72	8	206.57	0.721	204.74	0.717	7	180.30	0.920	7	178.85	0.919
230.82	-	-	-	-	-	-	-	-	-	-	-
246.57	-	-	-	-	-	9	241.27	0.620	9	240.83	0.626
EXP	**NUM3**			**NUM3M**		**NUM4**			**NUM4M**		
f [Hz]	**N [-]**	**f [Hz]**	**MAC [-]**	**f [Hz]**	**MAC [-]**	**N [-]**	**f [Hz]**	**MAC [-]**	**N [-]**	**f [Hz]**	**MAC [-]**
16.32	2	13.82	0.527	13.77	0.527	2	22.05	0.550	2	21.98	0.550
32.16	-	-	-	-	-	-	-	-	-	-	-
66.81	4	57.95	0.700	57.94	0.700	10	269.19	0.692	10	268.97	0.697
103.32	5	92.54	0.792	91.69	0.808	-	-	-	-	-	-
112.73	-	-	-	-	-	-	-	-	-	-	-
145.94	9	238.88	0.622	238.75	0.617	-	-	-	-	-	-
189.72	8	180.54	0.906	179.07	0.908	8	205.53	0.714	8	203.83	0.711
230.82	-	-	-	-	-	-	-	-	-	-	-
246.57	9	238.88	0.668	238.75	0.672	-	-	-	-	-	-

Considering the values of frequencies collected in Table 1, and lack of similarity for most of modal shapes, it was clear, that none of the proposed schemes correctly describe the examined model. Two new schemes were proposed with various rigidity of supports (see UPD1 and UPD2 on Fig. 4). The stiffness of supports were the purpose subject of model updating.

4.3 Optimization tools and objective functions

For optimization the genetic algorithms (GA) available in the MATLAB environment in the Global Optimization Toolbox [7] have been used. The variables of optimization were stiffness (k1, k2, k3, k4 presented in Fig. 4) and in the second case k1 stiffness (horizontal one on the right side) and the Young's modulus. Each time, despite the different number of updated parameters, the maximum generation number of 100 has been set as one of the criteria to stop the optimization process. The number of individuals in each population was 50.

During updating two object functions (OF1 and OF2) have been used:

$$OF1=\sum_{i=1}^{4}(f_{in}-f_{ie})^2 \quad , \tag{4}$$

$$OF2=\sum_{i=1}^{4}(1-MAC(i,i)) \quad , \tag{5}$$

where: f_{in} - frequency obtained in numerical simulations, f_{ie} - frequency measured during dynamic testing, MAC - coefficient describing similarity between numerical and experimental frequencies.

5 Optimization results

For the considered variants (various support schemes and different target functions - described in Paragraph 4), the model optimization/updating procedure was repeated twenty times, the results were analysed and averaged. A summary of the obtained results is given in Table 3. The application of the OF2 function in UPD1M case did not give a definite solution, therefore the results were not presented in the table.

Table 3. Comparison of experimental and obtained from optimization frequencies [Hz]

EXP		UPD1M			UPD2M					
		OF1			OF1			OF2		
		k1=5.984e17 Nm/rad k2=0.32 Nm/rad k3=4.339e17 N/m k4=4.726e17 N/m			k1=5.066e17N/m E=4.380e10 N/m2			k1= 9.782e17 N/m E=3.709e10N/m2		
N [-]	f [Hz]	f [Hz]	Δf [%]	MAC [-]	f [Hz]	Δf [%]	MAC [-]	f [Hz]	Δf [%]	MAC [-]
1	16.32	16.28	-0.24	0.52	13.69	-16.09	0.50	12.49	-23.49	0.50
2	32.16	32.96	-2.52	0.55	31.24	-2.84	0.58	28.48	-11.42	0.58
3	66.81	63.73	-4.61	0.65	62.10	-7.05	0.68	56.62	-15.25	0.68
4	103.3	107.1	3.65	0.52	112.9	9.31	0.49	103.0	-0.33	0.49
5	112.7	-	-	-	-	-	-	-	-	-
6	145.9	137.6	-5.69	0.24	140.2	-3.97	0.44	127.8	-12.44	0.44
7	189.7	192.6	1.53	0.76	197.6	4.14	0.92	180.2	-5.04	0.92
8	230.8	-	-	-	208.4	-9.72	0.01	190.01	-17.68	0.01
9	246.6	252.5	2.42	0.543	266.0	7.90	0.63	242.6	-1.62	0.62

6 Conclusions

Updating of the frequencies results in a completely different model than assumed: no symmetry in the supports and a significant contribution of anti-rotation constraints.

However the accuracy of obtained frequencies is satisfying. For UPD1M and OF1 the differences between corresponding frequencies does not exceed 6%.

The presented updating do not consider the initial imperfection as well as the deformation caused by static load. This can be one of the reason of discrepancy of the results. Another potential reasons can be the material heterogeneity or reinforcement asymmetry.

The puzzling issue is the lack of compatibility of the mode shapes between numerical and experimental model. The similarity of modal forms was measured by MAC coefficient, which collecting the differences between all measurement directions in one value.

Due to the practical difficulties (arch is demanding structure) of precise locating of accelerometers and the character of measurements (as mentioned before, the measurements took place in parallel to a larger research project and had to respect location of the other sensors) the theoretical and actual location of accelerometers could differ a bit. Unfortunately, the arch was objectively small structure, so the influence of inaccuracy of location of sensors can be significant.

Finding the causes of inaccuracy of measurements and improving the model should be the goal of further research, but unfortunately the arch has already been destroyed in static examination and potential correction of position of accelerometers is impossible. However the conclusions from presented investigation will be used in future examination of similar structures.

References

1. O. S. Salawu, Eng. Struct., **19**, 9 (1997)
2. M. Friswell, J.E. Mottershead, *Finite Element Model Updating in Structural Dynamics* **38**, 1-286 (1995)
3. B. Miller, CAMES, **18**, 191-203 (2011)
4. T. H. G. Megson, *Structural and Stress Analysis*, 130-145 (2014)
5. T. Uhl, W. Lisowski, *Praktyczne problemy analizy modalnej konstrukcji*, (1996)
6. T. Uhl, *Komputerowo wspomagana identyfikacja modeli konstrukcji mechanicznych*, (1997)
7. *Global Optimization Toolbox User's Guide, MATLAB R2018a,* (2018)

MATEC Web of Conferences **285**, 00011 (2019) https://doi.org/10.1051/matecconf/201928500011
DYNKON 2019

Dynamic problems in the operation of overresonant vibrating screens

Filip Matachowski[1] , and *Sławomir* Badura[1]

[1]AGH University of Science and Technology, 30-059 Kraków al.Mickiewicza 30, Poland

Abstract. The paper presents the results of the numerical analysis of the stress state (FEM) which is created in the construction elements of the PZ3R vibrating screen in normal operation conditions. These results where compared to and verified by tensometric, displacement and acceleration measurements. The results of strength analyses indicated the directions of possible modifications of structural elements of the screen, and also confirmed the desirability of their implementation, but in strictly specified order.

1 Introduction

Designing a structure is a difficult process, especially when its dynamic loads are not fully recognized, which in the next step does not allow for proper assessment of the stress in its elements.

It turns out that constructory experience is not enough for a newly designed structure to work properly, although it already has a predecessor in its class, the operation of which does not raise any objections [1].

This is the problem with the construction of the PZ3R vibrating screen, which was designed like a PZ screen, introducing changes, mainly dimensional ones, [2]. The elements of the new PZ3R screen construction began to crack after a relatively short period of device operation. The authors of the study decided to explain the reasons for this state of affairs, based on the results of analyzes of the stress state which arises in the construction elements of the vibrating screen in normal operation [3].

2 Analysis object

Designing a structure is a difficult process, especially when its dynamic loads are not fully recognized, which in the next step does not allow for proper assessment of the stress in its elements.

The PZ3R screen is not a classic resonance screen [4]. It differs from typical PZ screens by smaller dimensions of the sieve deck, a larger number of decks and an approximately parallel arrangement of vibrators to the sieve plates (Fig. 1).

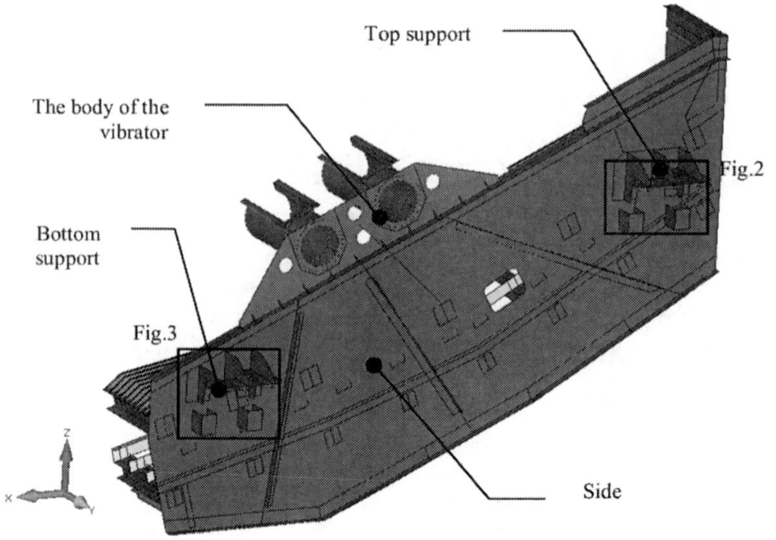

Fig. 1. Longitudinal section of the PZ3R screen - side view.

These changes have resulted in different kinematic characteristics of individual points along the length of the screen. There are vibrations from linear near the center of gravity of the vibrating screen to flattened ellipses of different inclination [5] near the supports. Three PZ3R screens were installed in one of the Polish mines for the preliminary classification of copper ore. During use, some structural elements began to break. Among others:
- cracking in the area of the props of the riddle side (Fig.2),
- permanent deformation of the support (Fig.2).

Fig. 2. Crack in the area of top support of side of the riddle (exact loaction shown on Fig.1), damage to HuckBolt screws (left picture). Permanent deformation of support (right picture).

Attempts have been made to restore the operating efficiency of the vibrating screen, by strengthening places particularly vulnerable to damage, which only temporarily reduced adverse phenomena. In order to comprehensively solve the problem, a mechanical (numerical) analysis of the screen's structural elements was performed, the results of which were verified by an experiment on the real object through tensometric measurement of strains (stresses) at the most intense points of the structure [5]. The results of these analyzes should indicate the directions of possible modernization of structural elements and, consequently, ensure full operational efficiency of the system.

Based on the results of the strength analysis [5,6], the user introduced a number of construction changes. These changes included (Fig. 3):

• Shortening the reach (distance from side to axis of the spring) of the support by 140mm

• Change in the structure of the side reinforcing ribs in the top and bottom support

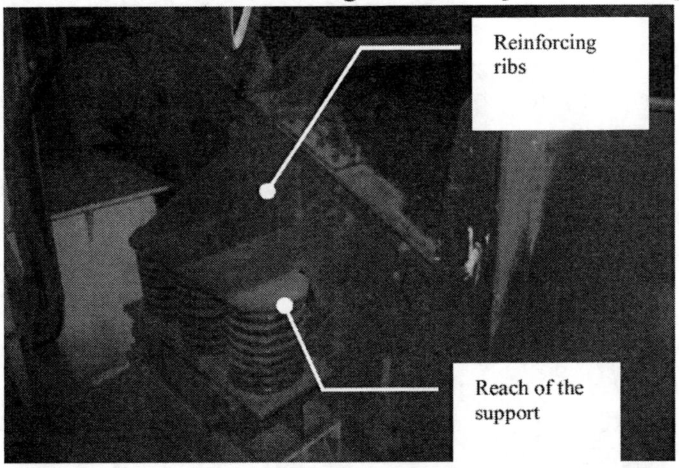

Fig. 3. View of the modified bottom support (see Fig.1 for location).

The conducted strength analysis after the changes [5] indicates the desirability of redesigning some elements of the screen, including supports. Strength analysis of the structure in which the reach of the support has been shortened by 140mm indicates that the local redesign of the support does not have any significant effect on reducing the stresses in the screen structure elements. Only the solution including additionally reinforced sides of the screen with C280 profiles results in significant reduction of stresses in the analyzed areas of the structure [5].

3 Analysis of the dynamics and kinematics of the vibrating screen

As part of these tests, it was performed:

• Traces of the screen vibrations trajectory in steady and transient states [7]

• Research on vibrator self-synchronization [8,9]

Moreover, in the field of sieve operation dynamics, the amplitude of fixed vibrations A and the required uplift index [4,10] were determined for proper operation of the vibrating screen in order to achieve the assumed capacity. The current values of these parameters are [1]:

• Vibration amplitude A=5.6mm against the required A=7mm

• The u_2=3.9 against the required u_2=5.88÷7.67 (u_2 – factor which define ratio of inertia force acting on single piece of material on vibrating screen to its gravity force)

It means that the dynamics of the screen in relation to the design assumptions has significantly decreased. In order to obtain the amplitude of vibrations A=7mm, the static moment of the vibrator should be increased. This change would cause the screen to operate at the maximum force, which would increase the strength problems of the structure [5]. Therefore, in relation to the construction of which the results of strength tests are described in [5], another structural change was introduced, consisting in changing the shape of the expansion beams connecting the ridding walls (Fig.1, 2 [5]). This change caused a further increase in the stiffness of the riddle. The research on the analysis of the dynamics and kinematics of the screen described in this study concerns a slightly different design, with a different stiffness than those described in the study [5].

4 The effect of structural changes on the stress state

The analysis of structural changes was made in such a way as to show the influence of each modification on the level of stress in the entire structure of the screen, which allowed to analyze its suitability.

All strength analyzes included in the study [5] were carried out taking into account the proposed modifications of the structure taking into account:
• Material changes
• Changes in support conditions
• Different load cases

The strength analyzes carried out in [5] showed a number of constructional irregularities causing concentration of stresses. For example, Figure 4 shows the concentration of stresses around the upper and lower supports.

Fig. 4. Visible areas of stress concentration around supports, at the lower thrust shaft and around deck beams.

5 Measurements of displacements, accelerations, deformations (stresses) of the actual structure of the PZ3R screen

The obtained results of the strength (numerical) analysis were verified by measuring strains (stresses) on the real object, in the most intense points of the screen structure elements selected as a result of the numerical analysis. In addition, measurements of displacements and accelerations of selected points of the screen structure were carried out.

5.1 Location of strain gauges on the construction of the PZ3R screen

Tensometric rosettes were placed around the location of the bottom support of the vibrating screen, three pieces on each side of the structure. Rosettes type "delta" No. 2 and

4 were mounted in the side part of the bottom support (Fig. 5). Rosettes "delta" type No. 1 and 3 were mounted in the bottom part of the lower supports of the vibrating screen. In the upper part of the lower legs, two-sensor rosettes No. 5 and 6 are mounted below support plate. Each measuring strain gauge was temperature compensated.

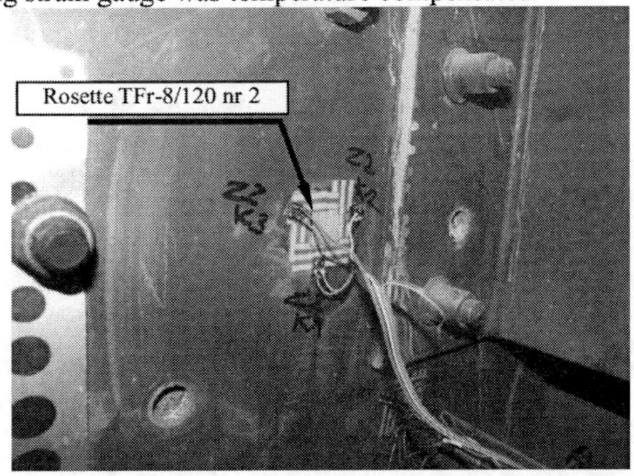

Fig. 5. Rosette No. 2 mounted in the side part of the bottom support of the vibrating screen.

During the experiment strains (stresses) of the above described points of the screen structure were made for the following phases of its operation [11]:

- Start-up,
- Steady work,
- Braking,
- Free run.

For example, in Fig. 6, the change of stress in all three phases of the screen operation registered by the No. 1 rosette, stuck - as numerical stress state analysis showed - in the area of the highest stress concentrations.

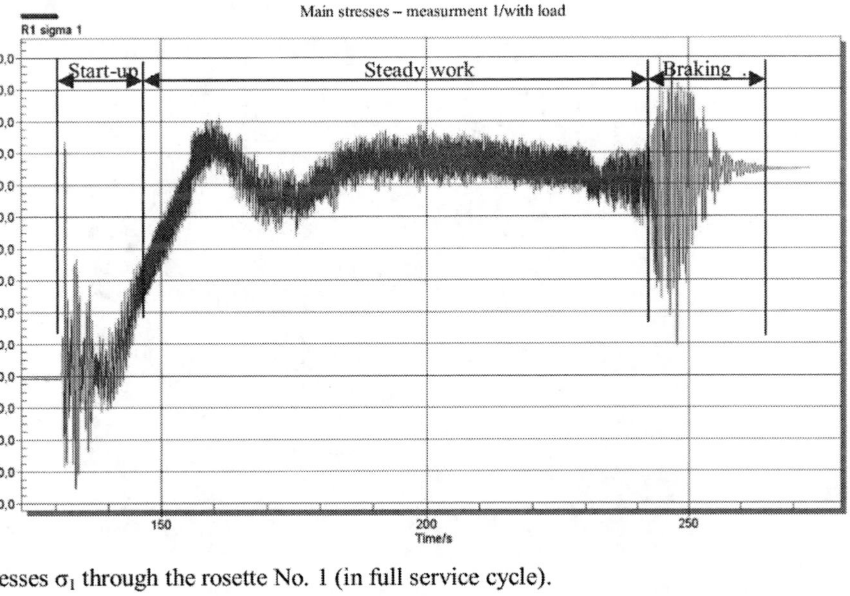

Fig. 6. Stresses σ_1 through the rosette No. 1 (in full service cycle).

The maximum stress values were indicated by the tensometric rosette No. 1 ($\sigma_{max}\cong$190MPa) during system braking (Fig. 6). At the same time, the strain gauge rosette no. 2 indicated stresses $\sigma max\cong$125MPa. The measurements were carried out after the modernization described in point 2 and the obtained stress values only slightly differed from the values of stresses in specific areas of the screen structure determined by numerical analysis (FEM).

5.2 Measurements of displacements

Displacement measurements were carried out using HBM WA 50 inductive sensors, connected to the MGCPlus bridge. During the measurements, displacements were recorded with three sensors mounted in three mutually perpendicular planes. The measuring instruments were mounted at the bottom left support. The WA50 sensors 1 measured the vertical displacements of the screen, the sensor WA 50 No. 2 measured longitudinal horizontal displacements and the sensor WA 50 No. 3 transverse horizontal displacements (Fig. 7 and 8).

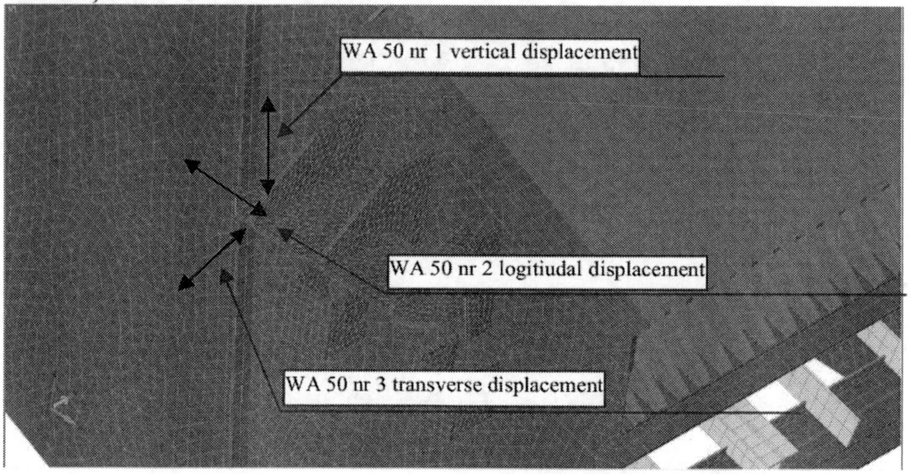

Fig. 7. Location of WA 50 displacement sensors on the left lower support of vibrating screen.

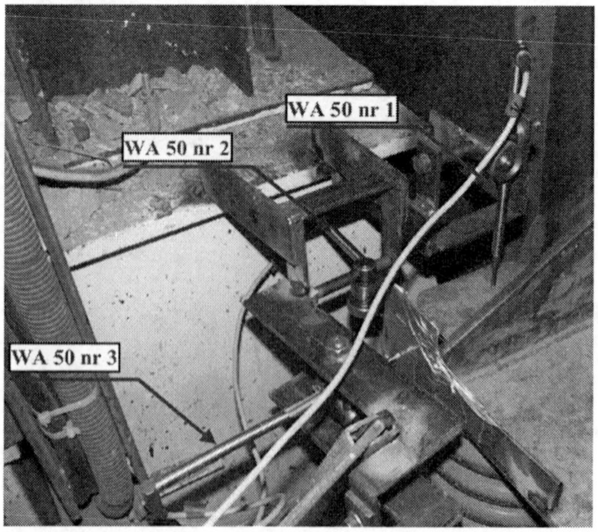

Fig. 8. WA 50 displacement sensors mounted on the vibrating screen.

The results of measurements of displacements of the indicated points of the screen structure elements, measured simultaneously with stresses, are shown in Figures 9, 10 and 11 respectively.

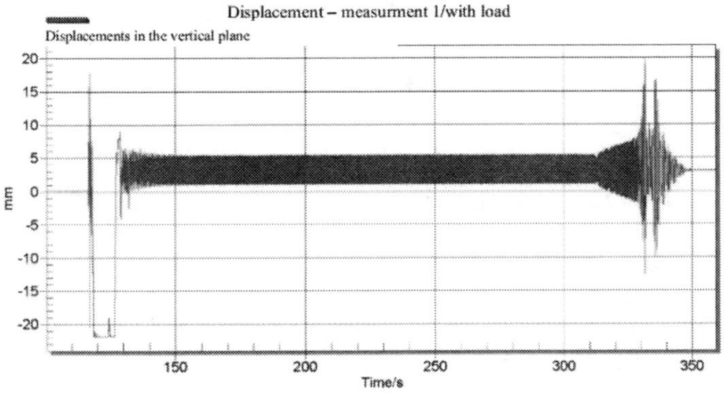

Fig. 9. Displacements in the vertical plane (with load).

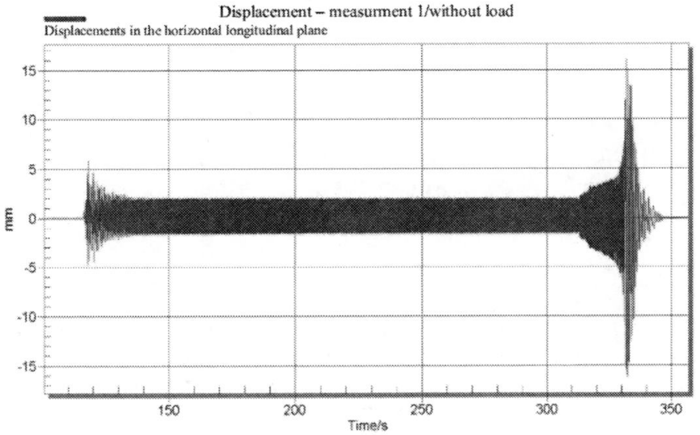

Fig. 10. Displacements in the horizontal longitudinal plane (without load).

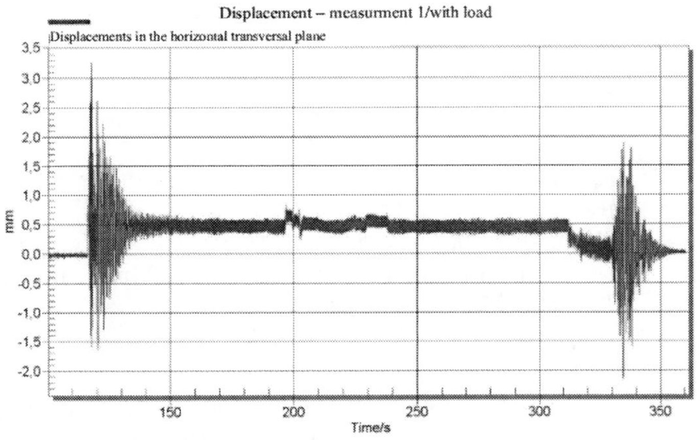

Fig. 11. Displacements in the horizontal transversal plane (with load).

The exact determination of the method of redesigning or making reinforcements in addition to the strength analysis requires the following tests:

- The correctness of self-synchronization in vibrators
- Vibration kinematics in characteristic points of the riddle
- Increase of vibration amplitude in transient states
- Observation of material flow on the sieve

• Acceleration of the riddle in steady as well as undefined states

Conducted and described research should be helpful in the development of changes in the structure and parameters of drive of screens [11].

6 Summary and Conclusions

The results of the strength analysis (FEM) of the PZ3R vibrating screen construction made for the obtained maximum system load values and the results of stress measurements in the construction elements of the actual screen allow the following conclusions:

1. The stress distribution in the construction of the PZ3R vibrating screen proves that due to the assumed fatigue life the structure has significant defects [5,12]. The effect of these defects is the concentration of stresses mainly in the areas of supporting supports (lower and upper) and the areas of vibrator construction.
2. The occurrence of technological and welding notches in the structure is the main cause of concentration of stresses
3. Reduction of stress in the most stressed places will be obtained as a result of structural changes:

• Installation of modified lower and upper supports
• Changes in the vibrator's construction
• Conversion of some nodes and reinforcement of structural elements

References

1. J.Raszyński, K.Szczepaniak, *Technical documentation. PZ3R – 2,2 x 6,0 vibrating screen*, Institute of Non-Ferrous Metals in Gliwice (2005)
2. J.Raszyński, K.Szczepaniak, *Technical drawings of PZ3R – 2,2x6,0 vibrating screen*, Institute of Non-Ferrous Metals in Gliwice (2005)
3. *Workflow of PZ3R screens. Position P3/4, P3/5, P3/6*, KGHM „Rudna" (2009)
4. T.Banaszewski, *Vibrating screens*, Publishing house „Śląsk" (1990)
5. Report on research work *Strength analysis of the PZ3R vibrating screen construction - 2,2, x 6,0*, KWZMiK, Kraków (2009), unpublished
6. Wolny S., Banaszewski T., Dzik S., Matachowski F, *Evaluation of the state of stress in structural components of a screen separator*, Mechanics and Control, vol.**29** no 2 s.88-96, (2010)
7. T.Banaszewski, A.Filipowicz, *Badania poprwaności drgań przesiewaczy*, Maszyny Górnicze nr.**90** Gliwice (2002)
8. T.Banaszewski, *Wpływ położenia samosynchronizujących się wibratorów na drgania przesiewacza*, Silesian University of Technology, Series - Mining nr.**245** (2000)
9. T.Banaszewski, *Wykorzystanie samosynchronizacji wibratorów do wzbudzania drgań przesiewaczy*, Scientific Seminar „Aplikacje badań naukowych w budowie maszyn"; AGH Kraków (2000)
10. J.Michalczyk, T.Banaszewski, *Oddziaływanie dynamiczne maszyn stosowanych w przeróbce surowców mineralnych*, Wydawnictwa AGH, Kraków (2006)
11. Report on research work *Analysis of the work dynamics of the PZ3R vibrating screen and development of guidelines for introducing structural changes*, KWZMiK WIMiR AGH, Kraków (2009) unpublished
12. Polish norm PN-90/B-03200 *„Konstrukcje stalowe. Obliczenia statyczne i projektowanie"*

MATEC Web of Conferences **285**, 00012 (2019) https://doi.org/10.1051/matecconf/201928500012
DYNKON 2019

Frequency optimisation of composite cylinder using an evolutionary algorithm and neural networks

Bartosz Miller[1,*] and *Leonard* Ziemiański[1]

[1]Rzeszów University of Technology, Faculty of Civil and Environmental Engineering and Architecture

Abstract. The paper deals with the optimisation of dynamic properties of a composite cantilever cylinder. The optimised parameters are both the fundamental natural frequency f_1 as well as the gap in frequency space around a selected external excitation force allowing to avoid the resonance phenomenon. The optimisation is performed using a novel approach combining particle swarm optimisation and artificial neural networks. The evolutionary algorithms are used to solve the optimisation problem with many local minima while neural networks are used to substitute time-consuming finite element calculations of the minimisation problem objective function.

1 Introduction

Multilayered fibre-reinforcement composite materials are nowadays widely applied in many engineering structures, before all — but not only — in aircraft and automotive manufacturing. The composite structures are durable and reliable, unfortunately, their dynamic behaviour is not as widely studied and tested as in case of structures made of traditional materials (e.g. steel). The slenderness and low mass of constructions made of multilayered composites may be the reason for a different behaviour of such structures and therefore should be carefully studied.

The problem studied in this work is related to the necessity to avoid resonance phenomenon in structures subjected to an external load with *a priori* known excitation frequency p. Resonance, appearing when the external excitation frequency p is equal (or very similar) to any of analysed structure's natural frequencies f_i, may be avoided either by maximisation of the fundamental natural frequency (assuming that the excitation frequency is low enough) or by creating a gap among all natural frequencies around the excitation frequency (when it is too high to fit below fundamental natural frequency).

The optimisation problem of maximisation of f_1 or creating the maximal gap around p is herein solved using one of the evolutionary algorithms, namely Particle Swarm Optimisation (PSO) (see [2]). The evolutionary algorithm has been chosen since the optimisation problem is non-convex, with a number of local minima, and may be rather difficult to be solved using traditional optimisation algorithms. Moreover, the values of the objective function being minimised are not computed using the usual approach involving Finite Element Method (FEM). The approach involving FEM, though very precise and reliable, is extremely time-consuming when using evolutionary algorithms (including PSO) that need to repeatedly calculate the

*e-mail: bartosz.miller@prz.edu.pl

value of the objective function. Artificial Neural Networks (ANN) (see [3, 4]) are applied instead of FEM since properly trained and tested ANN is able to give the objective function value almost instantly.

The proposed optimisation procedure uses the fact that in the multilayered composites structures it is possible to control, to a certain extent, the dynamic parameters by only changing the lamination angles (see [5]). Therefore, the control variables are the data describing the fibre angles in each of the composite layers (so-called *stacking sequence*), while the optimised values are either f_1 or the width of the gap around excitation frequency p.

2 The investigated model and the proposed optimisation procedure

2.1 The investigated model

The investigated model is a thin-walled cylinder fixed at one end, the length of the cylinder is $l = 6$ m, the diameter $R = 0.6103$ m and the wall thickness $t = 16$ mm. The cylinder walls are made from multilayered fibre-reinforcement composite material with the properties taken from [6]: $E_1 = 141.9$ GPa, $E_2 = 9.78$ GPa, $\nu_{12} = 0.42$, $G_{12} = 6.13$ GPa and $\rho = 1445$ kg/m^3. The finite element model of the thin-walled cylinder (see fig. 1) is built in commercial FEM code Adina [1] using multi-layered shell 4-node MITC4 elements (first order shear theory), the number of which equals 80 along the circumferential direction and 120 in the axial direction. The overall number of FE elements, nodes and degrees of freedom equals 6900, 9680 and 58080, respectively. The number of composite layers n in cylinder wall is equal to eight.

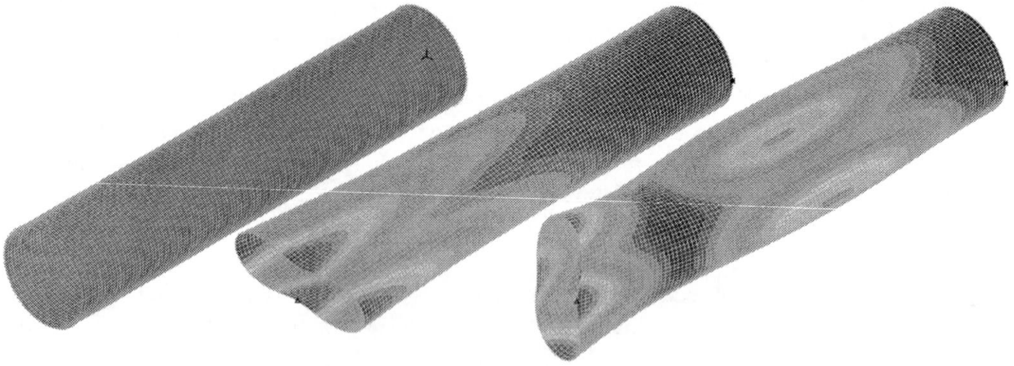

Figure 1. FE model of the thin-walled cylinder, the first and the fifth mode shapes

2.2 The optimisation problem

The optimisation tasks to be performed, namely the maximisation of the fundamental frequency f_1 or the maximisation of the gap around excitation frequency p, may be described as:

$$\mathbf{\Lambda}^* = \arg\min_{\mathbf{\Lambda}\in\mathbb{L}^8} \{\mathrm{g}(\mathbf{\Lambda})\} \ , \tag{1}$$

where $\mathbf{\Lambda}$ is a stacking sequence vector describing composite lamination angles $\mathbf{\Lambda} = \{\lambda_1, \lambda_2, \ldots, \lambda_8\}$, $\mathbb{L}^8$ is an 8-dimensional space of possible values of $\mathbf{\Lambda}$ and $\mathrm{g}(\mathbf{\Lambda})$ is an objective function to be minimised:

$$\mathrm{g}(\mathbf{\Lambda}) = -f_1(\mathbf{\Lambda}) \tag{2}$$

for maximisation of f_1, where $f_1(\mathbf{\Lambda})$ shows a relation between the fundamental frequency f_1 and the stacking sequence $\mathbf{\Lambda}$, and

$$g(\mathbf{\Lambda}) = -\min_{i=\{1,\ldots,10\}} |f_i(\mathbf{\Lambda}) - p| \tag{3}$$

for maximisation of the frequency gap around excitation frequency p (where $|\cdot|$ indicates the absolute value). The optimisation tasks discussed herein can be described by the following formulas:

$$\mathbf{\Lambda}^* = \arg\min_{\Lambda\in\mathbb{L}^8} \{-f_1(\mathbf{\Lambda})\} \tag{4}$$

for maximisation of f_1 and

$$\mathbf{\Lambda}^* = \arg\min_{\Lambda\in\mathbb{L}^8} \left\{-\min_{i=\{1,\ldots,10\}} |f_i(\mathbf{\Lambda}) - p|\right\} \tag{5}$$

for maximisation of the frequency gap around excitation frequency p. The values of the objective function are in this paper calculated using ANN, the Eq. 4 can be given in a form including the ANN application:

$$\mathbf{\Lambda}^* = \arg\min_{\Lambda\in\mathbb{L}^8} \{-\mathrm{ANN}_1(\mathbf{\Lambda})\} \tag{6}$$

Similarly, Eq 5 can be given as:

$$\mathbf{\Lambda}^* = \arg\min_{\Lambda\in\mathbb{L}^8} \left\{-\min_{i=\{1,\ldots,10\}} |\mathrm{ANN}_i(\mathbf{\Lambda}) - p|\right\} \tag{7}$$

where $\mathrm{ANN}_i(\mathbf{\Lambda})$ is the value of the ith natural frequency calculated by ANN for the lamination angles set $\mathbf{\Lambda}$.

To get an ANN properly approximating the FEM solution:

$$\mathrm{ANN}(\mathbf{\Lambda}) \cong \mathrm{FEM}(\mathbf{\Lambda}) \tag{8}$$

a number of vector pairs (called *patterns*), composed of $\mathbf{\Lambda}^j$ and $\mathbf{f}^j = \{f_1^j, f_2^j, \ldots, f_{10}^j\} = \mathrm{FEM}(\mathbf{\Lambda})$, have to be prepared. Most of the pairs $(\mathbf{\Lambda}^j, \mathbf{f}^j)$ create a learning set applied to tune the ANN parameters, the others create a testing set applied to check the accuracy of the trained ANN.

3 The results of the optimisation

The first step in the proposed procedure is to teach neural networks to predict the dynamic properties of the considered structure. Supervised learning of the ANN requires, in turn, the preparation of a set of patterns. The patterns are generated using FEM for different combinations of lamination angles, $12,664$ of them create the learning set and $8,000$ create the testing set. The overall number of FE calls necessary to create both pattern sets should be considered together with the following values:

- $20,664$ — the number of FE calls in case a novel PSO+ANN procedure described in this paper is applied,
- $160,000$ — the estimated number of FE calls in case classical optimisation procedure (with neither PSO nor ANN involved) is applied; the estimation is obtained on the basis of the number if FE calls during the maximisation of f_1 performed using classical approach,

- $2,800,000$ — the estimated number of FE calls during the PSO optimisation involving FEM (not ANN) to compute the objective function values; the estimation is obtained as a simple multiplication (the swarm size: 100; the average number of PSO iterations: 80; each procedure repeated 50 times; 7 optimisation cases: maximisation of f_1, maximisation the gap around $p = 30$ Hz, 40 Hz, 50 Hz, 60 Hz, 70 Hz and 80 Hz).

In case of PSO+FEM procedure, the legitimacy of using ANN is obvious because the number of FE calls is reduced by two orders of magnitude. In case of the application of classical methods, the number of FE calls is higher than for PSO+ANN procedure by about 8 times, the time consumption will increase significantly since the trained ANN gives its approximation almost instantly while FEM needs at least a dozen of seconds to calculate the desired result.

The neural networks are trained to give the values of the first ten natural frequencies of the composite cylinder for a given set of lamination angles which allows calculating the objective function values almost instantly (see Eq. 2 and Eq. 3). Once the ANN is ready the main PSO optimisation starts.

Table 1. Maximisation of the distance between any natural frequency f_i and excitation frequency p

External excitation frequency	Minimal distance to f_i absolute value	relative value
< 30 Hz	maximal f_1	= 44.99 Hz
30 Hz	12.9 Hz	23.6%
40 Hz	13.5 Hz	33.6%
50 Hz	20.8 Hz	41.5%
60 Hz	23.7 Hz	39.5%
70 Hz	24.0 Hz	34.3%
80 Hz	27.9 Hz	34.8%

The maximal value of f_1, obtained through PSO+ANN optimisation procedure, equals $f_1^{PSO} = 44.99$ Hz and corresponds to stacking sequence $\Lambda = [81.56/-37.65/15.22/-1.80/5.61/-1.91/-7,02]$. The optimisation of f_1 was performed using also classical optimisation approach, namely interior-point optimisation in Matlab environment, and the obtained result is $f_1 = 44.10$ Hz. The results obtained from the proposed PSO+ANN procedure is slightly better and, what is of great importance, obtained using a significantly smaller number of FE calls (and, of course, much faster). Table 1 shows the results of PSO+ANN maximisation of the distance between any natural frequency f_i and excitation force frequency p. For each of the considered values of excitation frequency, the distance to the nearest natural frequency is higher than 23% of p, what means that the procedure correctly selected the lamination angles and the resonance phenomenon will not appear.

Figure 2 shows the same results like the ones shown in Table 1 in a graphical form. Subsequent rows in Fig. 2 present the results of f_1 maximisation (the lowest one) and the maximisation of the distance between arbitrarily selected excitation frequencies (30 Hz through 80 Hz) and neighbouring natural frequencies. The green stripes represent the distance in a frequency space around selected excitation frequency (30 Hz through 80 Hz) free from natural frequencies, obtained when a particular set of lamination angles is applied. The thick vertical lines represent natural frequencies calculated for the same lamination angles. Although the results, as shown in Table 2, are reasonable from the point of view of resonance avoidance, the upper limit of the frequency gap obtained for $p = 30$ Hz smaller than the maximum f_1 obtained from f_1 maximisation, this result is incorrect since the stacking sequence obtained

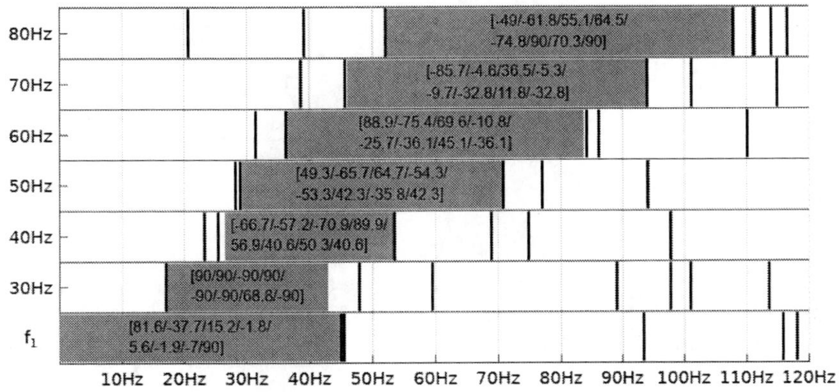

Figure 2. The frequency gap width for different external excitation frequencies

for f_1 maximisation gives better results (the frequency gap is wider than the one obtained for $p = 30$ Hz) so it is possible to find a better result in this case.

The neural network is thus additionally trained using the new patterns obtained during the first stage of the optimisation described above. Each repetition of the optimisation procedure for each optimisation case is finished with FE verification of the obtained results, the verification data (stacking sequence and corresponding FEM generated natural frequencies) may be treated as an additional example (pattern) to be used during ANN training. Altogether a set of $2,358$ patterns are obtained and the ANN is retrained using this pattern set combined with the previous learning pattern set. The overall number of FE calls is increased to $23,022$, what can be still treated as a very small part of the number of FE calls when the FE is used to calculate the objective function during PSO optimisation.

Table 2. Repeated maximisation of the distance between f_i and p

External excitation frequency	Minimal distance to f_i 1^{st}step	2nd step
< 30 Hz	$f_1^{max} = 44.99$ Hz	$f_1^{max} = 45.03$ Hz
30 Hz	12.9 Hz	15.0 Hz
40 Hz	13.5 Hz	13.3 Hz
50 Hz	20.8 Hz	20.9 Hz
60 Hz	23.7 Hz	24.2 Hz
70 Hz	24.0 Hz	24.5 Hz
80 Hz	27.9 Hz	28.3 Hz

Table 2 and Fig. 3 show the results of the PSO+ANN optimisation performed using retrained ANN, the maximal value of f_1 equals $f_1^{PSO} = 45.03$ Hz (corresponding stacking sequence is $\mathbf{\Lambda} = [-73.8/51.6/8.3/-13.3/4.8/2.7/-2.3/90]$). The improvement in f_1 is small, the frequency gaps around excitation frequencies do not raise such doubts as previously described for $p = 30$ Hz. The results in Fig. 3 show that the investigated $\mathbb{L}^8$ space of possible stacking sequences gives, from the point of view of f_1 and frequency gaps maximisation, a number of solutions similar when considering the result (f_1 or frequency gap width) but with different stacking sequence. It is clearly visible for f_1 maximisation and maximisation of the gap around $p = 30$ Hz, the dynamic properties of the structure are almost the same,

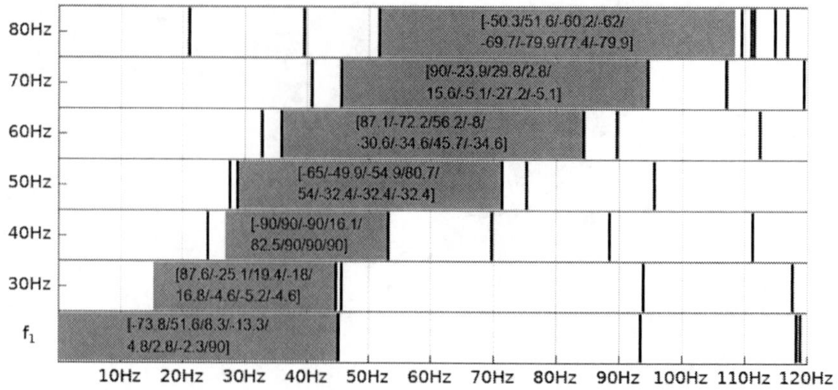

Figure 3. The frequency gap width for different external excitation frequencies (2nd step)

while stacking sequences are different ($\mathbf{\Lambda} = [-73.8/51.6/8.3/-13.3/4.8/2.8/-2.3/90]$ and $\mathbf{\Lambda} = [87.6/-25.1/19.4/-18/16.8/-4.6/-5.2/-4.6]$, respectively). Nevertheless, the proposed PSO+ANN procedure is able to find the expected solution.

4 Final remarks

The paper presents a PSO-based method of stacking sequence optimisation in laminated structures. The proposed PSO+ANN procedure can be applied to the optimisation cases where analytical formulae combining design variables and optimised quantities are not possible to derive. The proposed method, due to the replacement of FEM in the calculation of the dynamic parameters of the structure by ANN, accelerates the optimisation and allows to perform the optimisation in a limited time.

The combined PSO+ANN procedure has been successfully applied to problems related to avoiding vibration resonance, which is a major concern for every structure subjected to periodic external excitations. The presented examples show two approaches to the resonance avoidance, through maximisation of the fundamental natural frequency (for external excitation frequencies possibly smaller than the fundamental natural frequency) or through maximisation of a frequency gap around the external excitation frequencies (in other cases).

References

[1] Bathe Klaus-Jurgen, *ADINA: Theory and Modeling Guide. Volume I: ADINA Solids & Structures* (ADINA R&D, Watertown, 2016)

[2] Chopard Bastien, Tomassini Marco, *An Introduction to Metaheuristics for Optimization* (Springer, 2018)

[3] Haykin Simon, *Neural Networks and Learning Machines (3rd Edition)* (Pearson, 2008)

[4] Miller Bartosz, Ziemiański Leonard, Computer Assisted Mechanics and Engineering Sciences *7(4)* 781–793 (2000)

[5] Miller Bartosz, Ziemiański Lenoard, *Shell Structures: Theory and Applications Volume 4: Proceedings of the 11th International Conference "Shell Structures: Theory and Applications", (SSTA 2017), October 11-13, 2017, Gdańsk, Poland* (CRC Press, Boca Raton 2018)

[6] Vo Thuc Phuong, Lee Jaehong, Ahn Namshik, Composite Structures *89(4)* 524–535 (2009)

Arrival Traffic Separations Modelling By Using Continuous Probability Distributions

Adrian Pawełek[1,*] and *Piotr* Lichota[1,**]

[1]Institute of Aeronautics and Applied Mechanics, Faculty of Power and Aeronautical Engineering, Warsaw University of Technology, Nowowiejska 24, 00-665 Warsaw, Poland

Abstract.
This article presents a method that allows to analyze selected aspects of past arrival traffic by modelling distributions of time separations of arriving aircraft in a chosen navigation point of Terminal Manoeuvring Area with the use of continuous probability distributions. Modelling arriving aircraft time separations distribution with continuous probability density functions allows to apply various mathematical tools to analyze separations distributions. Moreover, by comparing distributions parameters, quantitative analysis of separations for days with various arrival traffic intensity can be performed. Assumptions, mathematical model, application in the exemplary experimental scenario with an airport and days with low and high traffic intensity, and results are presented in this article. Real air traffic data was used for the experimental scenario. Outcomes show that the method can be used for air traffic post-analysis, e.g assessment of maintaining separation.

1 INTRODUCTION

Constantly growing air traffic has a negative impact primarily on the environment, air operations safety, workload of air traffic controllers (ATC) and plane crew. Therefore, there is a need to seek for new methods which will allow for better and more robust analysis and organization of air traffic. A solution to these problems can be application of new air traffic sequencing methods. This will organize air traffic in an automatic and, at the same time, optimal way. Moreover, it will lower human workload, especially in departure and arrival phases that will result in a reduction of mental stress and its negative effects [1], [2]. Initial analysis showed that a solution for descent phase may be to combine Continuous Descend Operations (CDO) and Required Time of Arrival (RTA) assigned to each flight [3], [4]. Use of CDO allows to eliminate segments of horizontal flight. This reduces fuel consumption in aircraft descent phase, and thus lowers exhaust emission and flight cost. On the other hand RTAs allow to control trajectory in fourth dimension – time, what results in a better assessment of separations between aircraft. Therefore, it is less expected that radar-vectoring or go-around will be required due to too low separation. This concept could also contribute to the Aicraft Supervision System Concept [5] that assesses overall external and internal aircraft status.

*e-mail: apawelek@meil.pw.edu.pl
**e-mail: piotr.lichota@pw.edu.pl

Implementation of this approach requires detailed analysis of past air traffic and study of wide range of possible scenarios, with a focus on separations between aircraft. The approach based on time separations, so-called Time-Based Separation (TBS) [6] is also required for 4D trajectories application.

Modelling of arriving aircraft time separations distribution at a given waypoint of Terminal Manoeuvring Area (TMA) for past air traffic with the use of one of known probability distribution will allow to use wide range of mathematical tools for separations analysis. When distribution parameters are obtained from archival air traffic, new test cases can be automatically generated by sampling data from obtained probability distribution.

Current studies on air traffic analysis focus mainly on assessing the complexity of a current air situation and possibility of overloading of air sectors. They also classify ways to assess the complexity of air situation and methods used for these purposes, such as graph theory, dynamic systems theory, geometric approaches, and others. A detailed overview can be found in [7] and [8]. These works also present models that allow to estimate the workload of ATC services. In [9] authors present the complex metric allowing to assess a current air situation, taking into account measurement uncertainty. The assessment of the values of aircraft separations was made in [10]. The authors used a simple classification in which separation below the required value is considered as a conflict, and otherwise a lack of a conflict. The test scenario covered the French airspace. It is worth noting that in none of these works the concept of four-dimensional trajectories [6] was used, and all of these works focus mainly on an overall assessment of a state of the airspace, often taking into account hypothetical cases, solved using sophisticated mathematical methods. This paper presents a method that allows to easily assess the air situation in the final stage of descent, in TMA, where the most intensive air traffic is.

2 SEPARATIONS IN AVIATION

Separations in aviation are standardized by ICAO Doc 4444 [11] which contains procedures for air traffic management. Vertical, lateral and horizontal separations can be distinguished. In descent phase, after entering TMA, every aircraft follows strictly defined trajectory, therefore lateral and vertical separations are preserved. However, there is a significant possibility of horizontal separation being too small, what is called a conflict. More precisely, conflict appears when separation between two approaching aircraft, one after another, is smaller than required minimum. Separation is determined basing on comparison of positions reported by both aircraft or on comparison of reported time of flying over given waypoint. If surveillance systems are used, such as radars, ADS-B and others, minimum separation shall not be smaller than 5 NM. This value can be slightly reduced after fulfillment of conditions described in ICAO Doc 4444. Additional requirement is separation resulting from presence of wake vortex turbulence [12] caused by preceding aircraft, which can be very dangerous for the following aircraft. Turbulence value is often assessed with the use of numerical models [13], which usually express only a part of reality, so there is still some level of inaccuracy. Therefore, in this case aircraft are classified by their maximum take-off mass.

Increasingly, TBS concept is used instead of distance to assess horizontal separations. It is caused by the fact that TBS allows to achieve better synchronization of air traffic, due to fact that aircraft have different speed values during descent. This is not addressed while analyzing distance separation, while time separations reflect this. Moreover TBS allows for more effective air traffic control in windy conditions. When wind speed is significant, wake vortex turbulence disperses faster and this physical phenomena is time-based. Finally, TBS is part of 4D trajectory concept, thus all range of features offered by this concept may be used. This results in optimized trajectories and shorter flight times.

3 FITTING PROBABILITY DISTRIBUTION

Values of separation between pairs of aircraft flying behind each other are obtained as a result of random process, which can be described by one of known probability distributions. Usually, gamma distribution is used to determine probability of occurrence of the same event after given amount of time. However, often other similar distributions are also used, i.e. weibull and lognormal. All of mentioned probability distributions have two parameters referring to scale and shape of the probability density function. As sampled data is limited to some number of observations, and results are often inconclusive, sometimes it is hard to assess which of these distributions should be chosen, and decision about selection of the distribution is made basing on the knowledge of the process [14].

To check if an obtained set of observations $(x_1, x_2, ..., x_n)$ belongs to the population of given probability distribution $f(x, \boldsymbol{\phi})$ where $\boldsymbol{\phi}$ is parameters vector to be estimated, assessment is made by distribution fitting and evaluation of those outcomes.

Fitting probability distribution consists in finding appropriate function, which describes behavior of a random variable in as good way as possible. For a random variable with known probability distribution $f(x|\boldsymbol{\phi})$ and samples $(x_1, x_2, ..., x_n)$ unknown parameters vector $\boldsymbol{\phi}$ has to be estimated. To estimate unknown values of distributions parameters Maximum Likelihood Estimation (MLE) method [15–17] was used. This method requires defining likelihood function L for data samples (1):

$$L(x_1, x_2, ..., x_n|\boldsymbol{\phi}) = \prod_{i=1}^{n} f(x_i, \boldsymbol{\phi}) \tag{1}$$

The value of likelihood function provides information about probability of obtaining the same data sample, with an assumption that sampling will be from chosen probability distribution. Fitting probability distribution with the use of MLE can be presented on the example of the gamma probability distribution [18], defined as follows (2):

$$f(x|\alpha, \lambda) = \frac{\lambda^\alpha}{\Gamma(\alpha)} x^{\alpha-1} e^{-\lambda x}; \quad \Gamma(z) = \int_0^\infty x^{z-1} e^{-x} dx \tag{2}$$

where x is a random variable, α, λ are the gamma distribution parameters, respectively shape parameter and inverse scale parameter, and Γ stands for gamma function [18].

By combining (1) and (2) and denoting $L = L(x_1, x_2, ..., x_n|\alpha, \lambda)$, likelihood function for the gamma probability distribution takes the following form (3):

$$L = \prod_{i=1}^{n} \frac{\lambda^\alpha}{\Gamma(\alpha)} x_i^{\alpha-1} e^{-\lambda x_i} = \left(\frac{\lambda^\alpha}{\Gamma(\alpha)}\right)^n \left(\prod_{i=1}^{n} x_i\right)^{\alpha-1} e^{-\lambda \sum_{i=1}^{n} x_i} \tag{3}$$

Estimation of unknown parameters $\boldsymbol{\phi}$ can be done by finding such a $\hat{\boldsymbol{\phi}}$ which maximizes the likelihood function L. In practice, minimum of negative log-likelihood function is seeked (4):

$$\hat{\boldsymbol{\phi}} = \operatorname{argmin}\left(-\log(L(\alpha, \lambda|x_1, x_2, ..., x_n))\right) \tag{4}$$

To solve equation (4), analytic methods, like comparing partial derivatives to zero, may be used, but for more complicated forms of the likelihood function iterative methods are selected. Initial estimate and, at the same time, a starting point for maximum likelihood method may be obtained by using method of moments. By using this method, initial estimates of shape and scale parameter can be calculated using the following equations (5):

$$\hat{\alpha} = \frac{\overline{x^2}}{\sigma^2}; \quad \hat{\lambda} = \frac{\overline{x}}{\sigma^2} \tag{5}$$

where $\overline{x}$ is the mean and σ^2 is the variance.

4 EXPERIMENTAL SCENARIO

Archival air traffic data was obtained from the Eurocontrol Data Demand Repository 2 (DDR2) database [19]. DDR2 contains, among others, planned and archival air traffic for European airspace. The data is stored in Aeronautical Information Regulation And Control [20] files that contain air traffic for 28 consecutive days. To obtain data from the DDR2, the Eurocontrol's software, Network Strategic Tool was used. Then, data on incoming traffic were exported. In this way, a set of two trajectories was obtained for each day: m1, the trajectory reflecting a last filled flight plan and m3, the trajectory obtained on a basis of radar measurements, determined after the flight. By comparing descent trajectories for m1 and m3, a decision was made that m3 data will be used for further analysis, as these are real trajectories that may differ significantly from planned ones, especially in the descent phase. In addition, they allow calculating the real distribution of the flight separation values.

The test scenario was prepared for Barcelona El-Prat airport (Spain), ICAO code: LEBL. The airport is known as one of the most busiest airport in Europe. The airport has three runways and five approach directions. The airport chart [21] is presented in Figure 1. The direction from which the plane will approach depends on a current configuration of the airport, which depends on a direction of wind and a time of a day. For Barcelona El-Prat airport, as defined in the Standard Arrival Chart (STAR), three airport operational configurations are possible. The most commonly used is Conventional / West Configuration, which corresponds to the STAR3 descent procedure [22]. The descent procedure can be started from eleven directions, which then are grouped in four Initial Approach Fixes (IAFs), where instrument approach begins. Then, from four IAFs, the air traffic is entirely directed to the Intermediate Fix (IF) (this is the navigation point at which the aircraft finished descent and starts the ILS procedure for landing), that is shown in the Instrument Approach Chart (IAC) [23] for this configuration (Figure 2). For this configuration, the direction is RWY 25R, for which IF is TEBLA. Due to the fact that in this point all aircraft arriving at the Barcelona El-Prat airport are grouped, this point was selected for the analysis of the separation values.

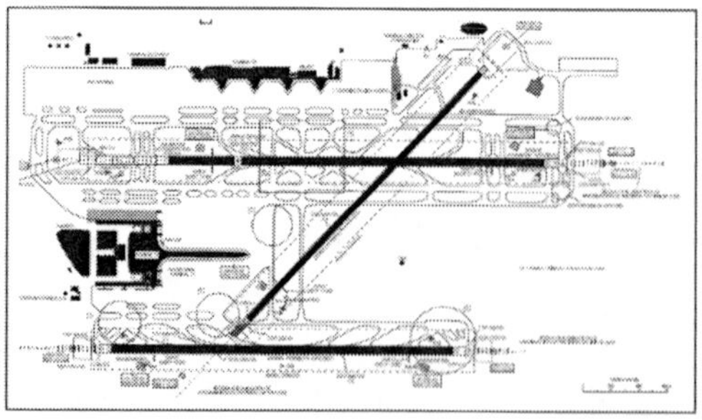

Figure 1. Diagram of Barcelona El-Prat airport (LEBL) [21]

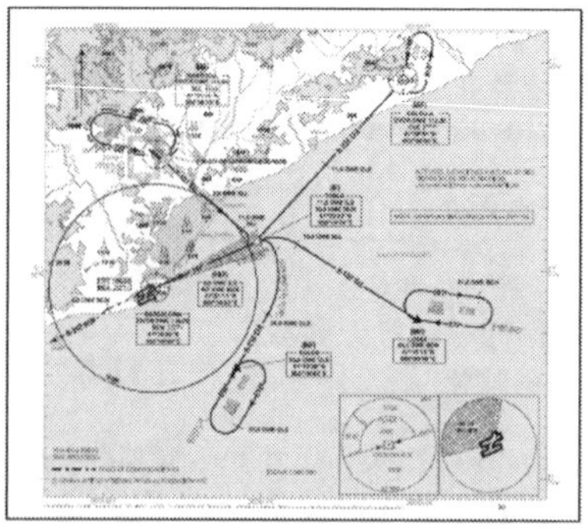

Figure 2. Instrument Approach Chart for RWY 25R [23]

To represent distribution fitting, two days were selected from the analyzed data set. During those days all planes followed the STAR3 procedure and the RWY 25R direction. Selected days: January 27th, 2016, a day with low arrival traffic - 229 landings took place during that day and April 3rd, 2016, a day with high arrival traffic - 380 landings took place during that day. For each pair of aircraft flying one after another, the time separation values were determined at TEBLA, by calculating absolute difference between times at which each of aircraft passed TEBLA fix. In that way time separations for whole day were obtained and were used as an input data for distribution fitting. Eighty-nine continuous probability distributions [24] were fitted to the data and distribution with the lowest value of Sum of Square of Errors (SSE) was chosen as the best fit.

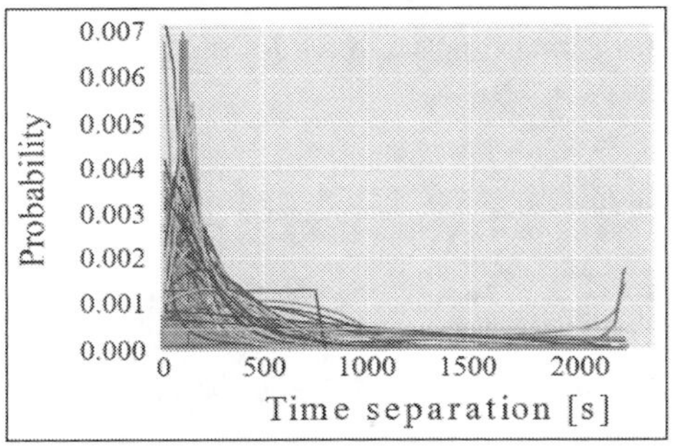

Figure 3. All fitted distributions for January 27th, 2016 (LEBL, IF=TEBLA)

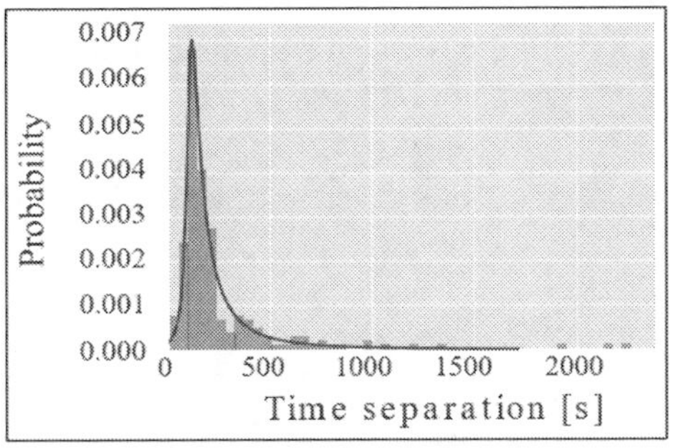

Figure 4. The best fitted distribution for January 27th, 2016 (LEBL, IF=TEBLA)

Figure 3 presents results of fitting all distributions for January 27th, 2016. The horizontal axis shows the value of time separation in seconds and the vertical axis shows probability value. The histogram bars present actual separations data, while the continuous lines present fits for different probability distributions. Results presented in Figure 3 show that only small amount of distributions among tested is close to reflect characteristics of time separations distribution.

Figure 4 presents the best fitted distribution for January 27th, 2016. The horizontal axis shows the value of time separation in seconds and the vertical axis shows probability value. The histogram bars present actual separations data, while the continuous line presents best fit probability distribution. Results presented on Figure 4 show that the distribution which fits best to the data is Johnson's S_U with shape parameters a=−0.93, b=0.69, location parameter loc=102.46 and scale parameter $scale$=29.63. The SSE for presented fit is $1.49 \cdot 10^{-6}$.

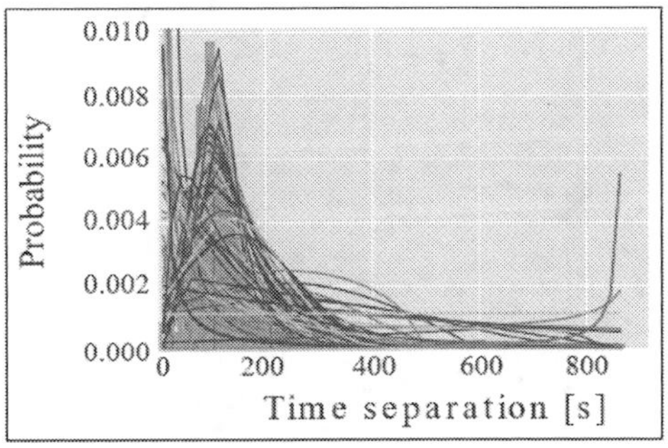

Figure 5. All fitted distributions for April 3rd, 2016 (LEBL, IF=TEBLA)

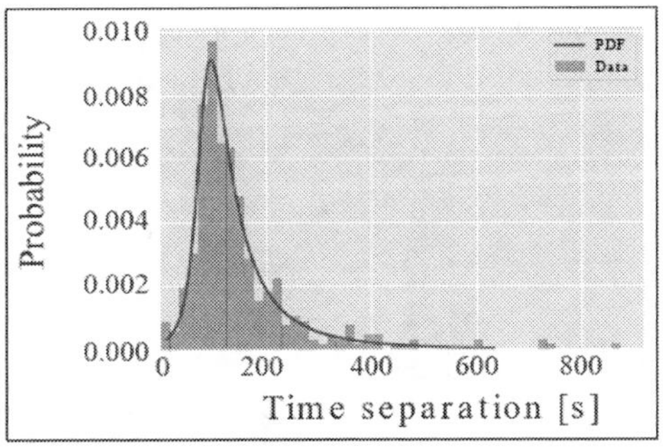

Figure 6. The best fitted distribution for April 3rd, 2016 (LEBL, IF=TEBLA)

Figure 5 presents results of fitting all distributions for April 3rd, 2016. Markings are analogous to Figure 3. Similarly to Figure 3, results presented in Figure 5 show that only some distributions fits were close to the data distribution.

Figure 6 presents the best fitted distribution for April 3rd, 2016. Markings are analogous to Figure 4. Results presented in Figure 6 show that the distribution which fits best to the data is Johnson's S_U with shape parameters $a=-1.65$, $b=0.83$, location parameter $loc=70.47$ and scale parameter $scale=22.41$. The SSE for presented fit is $6.17 \cdot 10^{-6}$.

Further analysis that consisted of ten days in which all planes followed the STAR3 procedure and the RWY 25R direction were performed. In eight of ten analyzed days the best fit was for the Johnson's S_U distribution, whilst for remaining two days Johnson's S_U was 2nd and 5th best fit (Table 1).

Table 1. Best fits for analyzed days

Day	Distribution* (SSE)				
1	J (3.82e-6)	L (7.64e-6)	G (1.13e-5)	D (1.43e-5)	N (1.94e-5)
2	J (1.78e-6)	L (6.45e-6)	B (7.97e-6)	H (1.01e-5)	I (1.01e-5)
3	J (6.70e-7)	L (9.00e-7)	A (1.53e-6)	B (1.89e-6)	H (1.90e-6)
4	J (2.22e-6)	L (5.89e-6)	A (6.69e-6)	B (6.74e-6)	K (6.74e-6)
5	J (1.49e-6)	G (3.91e-6)	F (5.20e-6)	C (5.40e-6)	L (7.80e-6)
6	J (9.38e-6)	G (1.32e-5)	L (1.34e-5)	B (1.46e-5)	K (1.46e-5)
7	J (6.16e-6)	L (6.35e-6)	N (1.58e-5)	C (1.63e-5)	M (1.66e-5)
8	J (6.17e-6)	K (1.71e-5)	B (1.72e-5)	L (1.74e-5)	E (1.93e-5)
9	F (1.11e-5)	C (1.29e-5)	N (1.54e-5)	M (1.65e-5)	J (1.86e-5)
10	L (6.12e-7)	J (1.03e-6)	B (1.25e-6)	H (1.70e-6)	I (1.70e-6)

* Distributions marking: alpha (A), Burr (B), Cauchy (C), double Weibull (D), Fisk (E), folded Cauchy (F), generalized normal (G), generalized extreme value (H), inverse Weibull (I), Johnson's S_U (J), Mielke (K), non-central t (L), t (M), Tukey lambda (N).

Linear regression was applied to obtained estimates of every of the Johnson's S_U distribution parameters versus number of arrivals in a given day. Results are presented in Figure 7. In all of the inset figures obtained estimates of parameters are marked with dots and linear regression is presented with the continuous lines. Horizontal axes present number of arriving aircraft, while vertical axes present parameter values.

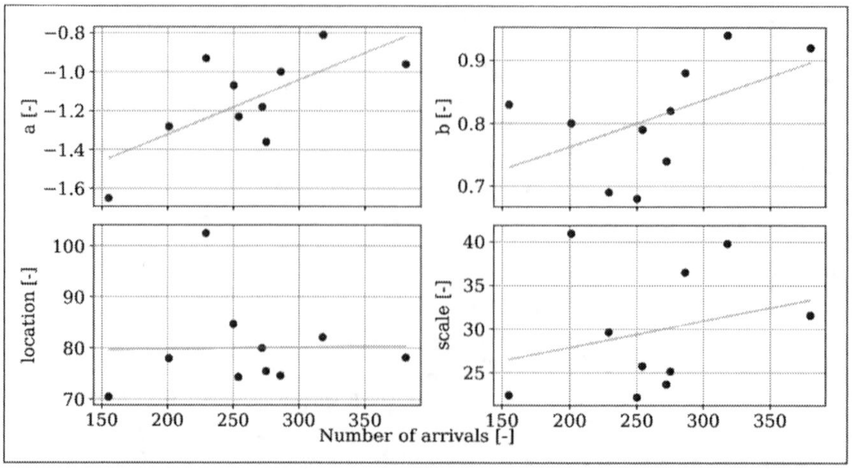

Figure 7. Linear regression for estimated values of the Johnson's S_U distribution parameters

From the plots in Figure 7, it can be observed that shape parameters a and b, as well as a scale parameter are increasing with air traffic density, whilst a location parameter is independent of air traffic density. From the regression analysis it is also possible to determine the separation distribution parameters for any traffic intensity. However, this requires more data points (e.g in order to select polynomial order) and is a subject of a separate study.

Results showed that regardless of the volume of air traffic only few of dozens distributions can be used to fit distribution of time separations, while the best fit can be achieved with use of the Johnson's S_U distribution. By comparing the results obtained for different air traffic intensities, it can be observed that shape and scale parameters change, while location parameter is approximately constant. It is due to the fact that for a smaller intensity of air traffic distributions are slightly more flattened, while the maximums of the functions do not move significantly.

5 CONCLUSIONS

In this article method for modelling of arrival traffic separations distribution for a chosen waypoint with the use of continuous probability distribution was presented. Realistic experimental scenario was prepared and presented method was applied, taking into account assumptions. Results showed that presented method can be used successfully to model separations distribution.

In the future more sophisticated methods may be considered to achieve more accurate models, especially for parameters regression. Moreover, creation of generalized model can be considered, which will allow to estimate distribution parameters as a function of a number of flights, airport configuration, wind speed and other arguments.

ACKNOWLEDGMENT

Authors would like to thank Faculty of Power and Aeronautical Engineering for funding this research as a part of the grant "Dynamic synchronization of arrival traffic for minimum-fuel trajectories".

References

[1] A. Kopyt, M. Biernacki, M. Piotrowski, *The study on the mental stress impact on pilots on the PZL SW-4 flight simulator*, Transactions of the Institute of Aviation, **32** (2014)

[2] A. Kopyt, P. Bartuzi, K. Zużewicz, *Experimental Identification of the Mathematical Model of Human-Operator Acting Under Mental Stress*, Acta of Bioengineering and Biomechanics **19**, 177 (2017)

[3] R. Dalmau, X. Prats, *Assessment of the feasible CTA windows for efficient spacing with energy-neutral CDO*, 7th Proceedings of the International Conference on Research in Air Transportation (2016)

[4] A. Pawełek, R. Dalmau, P. Lichota, X. Prats, *Arrival traffic synchronisation with Required Time of Arrivals for fuel-efficient trajectories*, Proceedings of the 17th ATIO-AIAA Aviation Technology, Integration, and Operations Conference pp. 1–14 (2017)

[5] J. Narkiewicz, M.Z. ans S. Topczewski, A. Kopyt, *A concept of Aircraft Status Supervision System*, Journal of Aerospace Engineering **30** (2017)

[6] L. Stridsman, *SESAR Concept of Operations Step 1*, Tech. Rep. D65-011, SESAR, Avenue de Cortenbergh 100, B-1000 Bruxelles (2012)

[7] D. Delahaye, S. Puechmorel, *Modelling and Optimization of Air Traffic* (Wiley-ISTE, London, UK, 2013)

[8] M. Prandini, L. Piroddi, S. Puechmorel, S.L. Brázdilová, *Toward Air Traffic Complexity Assessment in New Generation Air Traffic Management Systems*, IEEE Transactions On Intelligent Transportation Systems **12**, 809 (2011)

[9] M. Prandini, V. Putta, J. Hu, *A probabilistic measure of air traffic complexity in three-dimensional airspace*, International Journal Of Adaptive Control And Signal Processing **24**, 10 (2010)

[10] D. Gianazza, N. Durand, *Assessment of the 3D-separation of Air Traffic Flows*, ATM 2005, 6th USA/ Europe Air Traffic Management Research and Development Seminar (Baltimore, USA, 2005),

[11] ICAO, *Doc 4444, Procedures for air navigation services: Air Traffic Management*, ICAO, Montréal, QC, Canada, 16th edn. (2016)

[12] T. Gerz, F. Holzäpfel, W. Bryant, F. Köpp, M. Frech, A. Tafferner, G. Winckelmans, *Research towards a wake-vortex advisory system for optimal aircraft spacing*, Comptes Rendus Physique **6**, 501 (2005)

[13] K. Rogowski, *Numerical studies on two turbulence models and a laminar model for aerodynamics of a vertical-axis wind turbine*, Journal of Mechanical Science and Technology **32**, 2079–2088 (2018)

[14] Siswadi, C.P. Quesenberry, *Selecting among weibull, lognormal and gamma distributions using complete and censored samples*, Naval Research Logistics **29**, 557 (1982)

[15] V. Ricci (2005), *Fitting distributions with R, Release 0.4-21*, accessed: 21 Oct 2018, `https://cran.r-project.org/doc/contrib/Ricci-distributions-en.pdf`

[16] P. Lichota, *Inclusion of the D-optimality in multisine manoeuvre design for aircraft parameter estimation*,, Journal of Theoretical and Applied Mechanics **54**, 87 (2016)

[17] P. Lichota, J. Szulczyk, D. Agudelo Noreña, F.A. Vallejo Monsalve, *Power Spectrum Optimization In The Design Of Multisine Manoeuvre For Identification Purposes*, Journal of Theoretical and Applied Mechanics **55**, 1193 (2017)

[18] R.V. Hogg, A.T. Craig, *Introduction to Mathematical Statistics*, 4th edn (Macmillan, New York, US, 1978)

[19] Eurocontrol, *DDR2 Reference Manual For Generic Users 2.1.2*, Eurocontrol, Brussles, Belgium (2015)

[20] Eurocontrol, *Eurocontrol specification for the Electronic Aeronautical Information Publication (eAIP)*, Eurocontrol, Brussles, Belgium (2015)

[21] *Barcelona/El Prat plano de aeródromo-oaci*, AIP España (2016)

[22] *Barcelona/El Prat AD Standard Instrument Arrivals (STAR) Runway 25R/25L*, AIP España (2013)

[23] *Carta de aproximación por instrumentos Barcelona/ El Prat ILS Z RWY 25R*, AIP España (2016)

[24] E. Jones, T. Oliphant and others, *SciPy: Open source scientific tools for Python*, http://www.scipy.org/, (2001–), Accessed: 2019-03-31

Numerical model of tracked vehicle crossing inland water

Mariusz Pawlak*[1], *Sebastian* Sławski[1], and *Sławomir* Duda[1]

[1]Silesian University of Technology, Institute of Theoretical and Applied Mechanics,

Konarskiego 18a., 44-100 Gliwice

Abstract. In this paper, a numerical model of tracked vehicle with a simplified suspension and a winch forcing movement at a constant speed of 2 m / s is presented. The simplified vehicle body was created in the CAD environment, the suspension model, dry and wet obstacle, as well as the winch system and material models, were created in the LS PREPOST. The body of the vehicle was treated as a rigid, mounted by springs and dampers on tracks, the solid element was replaced by a shell element with assigned moments of inertia and mass in the node located in the centre of gravity of the vehicle. The characteristics of the resultant contact force of vehicle travel on the rigid ground for a dry and water obstacle are presented.

1 Introduction

In the case of tracked vehicles, especially military ones, there is a need to cross terrain obstacles, for example, ditches, river beds, both dry and filled with water. In the case of a water obstacle, there is a risk of losing the vehicle's contact with the ground, which may cause a loss of control over the vehicle by the flow of the river. To be able to carry out the analysis of passing the tracked vehicle, a vehicle model with a modelled suspension, a definition of contact between the tracks and the ground, and modelling of motion are needed. For this purpose, the article presents a model with a simplified suspension and a winch forcing movement at a constant speed of 2 m / s. The LS DYNA software is an environment that enables the modelling of a vehicle suspension, with a forced movement of a rigid body as well as the modelling of the fluid-structure interaction. The simplified vehicle body was created in the CAD environment, the suspension model, dry and wet obstacle, as well as the winch system and material models, were created in the LS PREPOST. Because the body of the vehicle can be treated as a rigid, mounted by springs and silencers on tracks, the question was whether the solid element could be replaced by a shell element with assigned moments of inertia and mass in the node located in the centre of gravity of the vehicle. The characteristics of the resultant contact force of vehicle travel on the rigid ground for a dry and water obstacle will be presented.

* Corresponding author: mariusz.pawlak@polsl.pl

2 Comparison of the shell and solid model in the numerical buoyancy analysis

To compare the solid model with the shell model were created three cuboids with the same moments of inertia and mass. The models are presented in table 1. For calculation the following input data was adopted: length =5m along X-axis, width=4m along the Z axis and height=3m along Y axis. Assuming that density of water is ρ_w = 1000 kg/m^3, analysed object is heavier than water and with an averaged density of the solid ρ_b = 1000 kg/m^3, the total mass of solid will be $m = \rho_b \cdot V$ kg = 72000 kg. Moments of inertia in centre of gravity of a cuboid are as follow:

$$I_l=1/12 \cdot m(w^2+h^2) \qquad I_w=1/12 \cdot m(l^2+h^2) \qquad I_h=1/12 \cdot m(l^2+w^2) \tag{1}$$

For the coordinate system applied in model moment are as follow:

I_x = 150000 kg·m^2 $\quad$ I_y = 246000 kg·m^2 $\quad$ I_z = 204000 kg·m^2

Is assumed that centre of gravity is in the centroid of the cuboid (material is homogeneous) Xc=3.5m, Yc=2m, Zc=1.8m.

Table 1. Models description.

Model_1 Cuboid as solid body (l)5mx(h)3mx(w) 4m	Model_2 Cuboid as solid body with Part_Inertia	Model_3 Cuboid as a surface body with part inertia
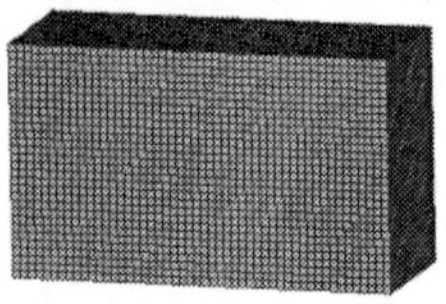	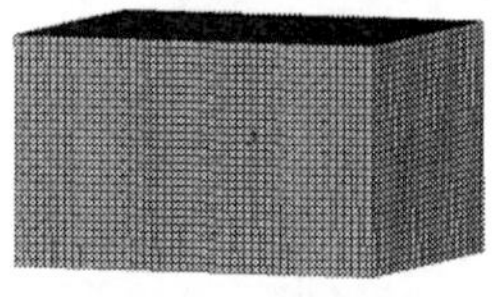	
126102 nodes, 60000 solids as rigid elements	126103 nodes, 6000 solids as rigid elements, 1 constrained extra_nodes_node	70684 nodes, 9400 shells as rigid elements 1 constrained extra_nodes_node

From the dynamic equation of translational motion

$$F = m \cdot a \tag{2}$$

Where mass is 48000 kg, and force F is the vector sum of the weight of the object and the buoyancy of the water.

$$F = F_c + F_w \tag{3}$$

$$F_c = m \cdot g \tag{4}$$

$$F_w = \rho_{water} \cdot g \cdot V \tag{5}$$

Forces in applied model are as follow:

F_c = 709920 N $\quad$ F_w = 588360 N $\quad$ F = 121560 N

Based on the difference in the density of water and solid, can be determined the acceleration of the falling of the body in the water:

$$a = g \cdot (1 - \rho_{water} / \rho_b) \tag{6}$$

Assuming the initial velocity V_0 = 0 m/s ,the path s_0 = 0 from the uniform motion equation $s = at^2/2$ and the acceleration of the body in the water 1.634 m/s^2, is determined the maximum falling time of the cuboid t = 2.47 s. There are many publications describing drop tests of bodies of various shapes into the water [6-12], and in derived theoretical models the resistance force resulting from the shape and the type of surface of the object is added to the buoyancy force. Based on those forces the resistance factors are calculated. In the presented drop tests were compared kinematic quantities to determine whether the model simplification by adopting only the outer layer of the coating with the assigned moments and the mass in the centre of gravity by part inertia command in LS Dyna will be sufficient. The characteristics of the drop test in the water for all three variants are presented. In the system is taken into account gravitational acceleration g = 9.806 m/s^2 parallel to the y-axis. Contact between the bottom plate located above the water border and the rectangular prism was established. Prepared model and hydrostatic pressure distribution were shown on Fig. 1.

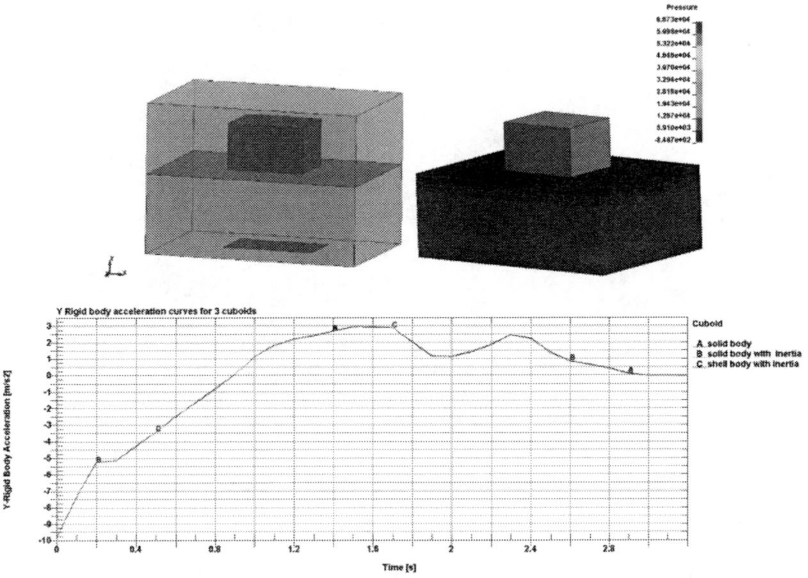

Fig. 1. Hydrostatic pressure distribution in the water and Y axis rigid body acceleration A)rigid solid body B)rigid solid body with mass and inertia assigned in the node located in gravity centre C)rigid shell body with mass and inertia assigned in the node located in gravity centre.

As can be noticed there is no difference in rigid body acceleration for all three cuboids, the difference between analytical calculation and numerical results of max. Acceleration is because the centre of the rigid body initially was in the air and initial acceleration was 9.806 m/s2 what resulted in the slightly larger time of drop test. Due to the size of the analysed system, the volume of water and air was discretized using the finite element method with regular cubic elements with a size of 0.1 m. The ALE grid is the same in the whole system, consisting of 1409860 nodes, 137200 solid deformable elements. An atmospheric pressure of 101,300 Pa has been set for the air, which can be seen in the form of pressure on the water surface. Along with depth, the value of hydrostatic pressure is increasing. The pictures below (Fig. 2) show the steps of immersing the object in water in chronological order.

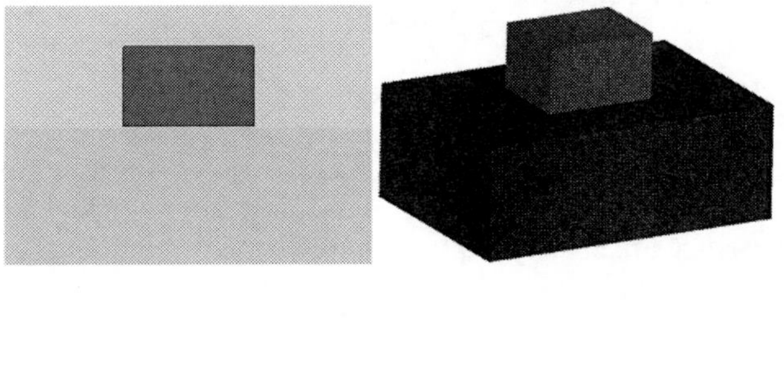

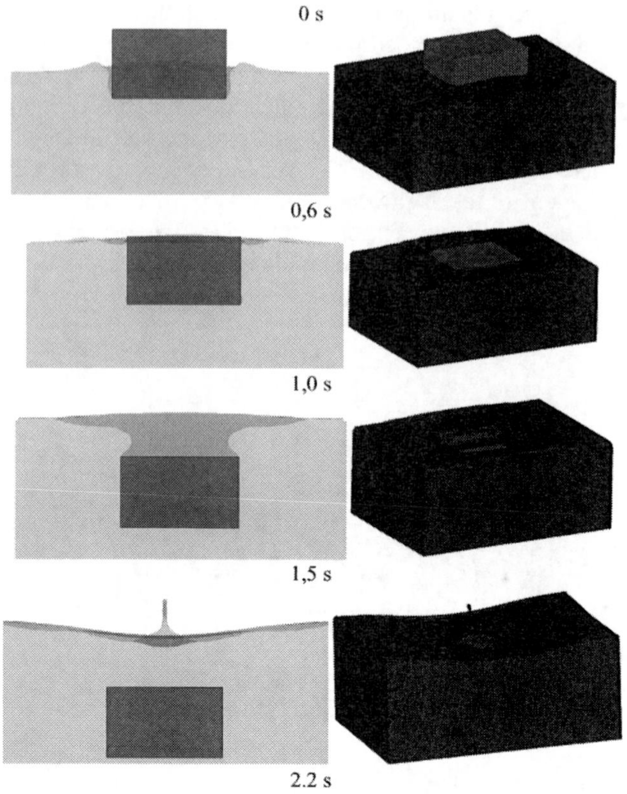

Fig. 2. Numerical results from Cuboid drop test in a function of time.

During the experimental drop test of cuboid into the water (Fig. 3), a higher amount of air bubbles trapped in the water was observed.

Fig. 3. Sample images from the real drop test of cuboid

Analysing the characteristics presented in Fig.1 can be noticed the compatibility between models. Due to the significant reduction in the number of finite elements in the model no. 3 and a significant reduction in simulation time, this method will be applied to the vehicle model with suspension.

3 The numerical model of the tracked vehicle

To analyse the dynamics of the tracked vehicle crossing through the water obstacle, the LS DYNA computing environment was selected. It is a software package for the analysis of fast-changing phenomena using the explicit finite element method to calculate engineering problems. The calculation system consists of a vehicle model with assigned mass moments of inertia, springs, dampers and tracks that simulate caterpillar tracks. As the first, analysis of dropping the rectangular prism from the height just above the water level to the bottom of the 5m deep reservoir was carried out. It was necessary to investigate whether there is a difference between a rigid solid model and a rigid shell model with assigned moments of inertia and mass in the node. The number of finite elements was significantly reduced due to the use of the rigid method and the shell model. The second test was the simulation of passing through a dry obstacle and the results presented in Chapter 4 This is due to the necessity of checking whether the object moves with a constant maximum linear speed of 2m / s to avoid a greater dynamic pressure distribution during fluctuating acceleration and deceleration in water. The next stage was the creation of a water environment using elements of ALE (Euler Lagrange Equations) used in modelling strongly non-linear phenomena, such as, for example, explosions in air and water. Due to the large size of the facility, along with the obstacle, there were restrictions on the minimum finite element sizes, for which the computing power of the computer will be sufficient. Simulations were carried out in the LS DYNA 9.0 environment. The CAD model from the Autodesk Inventor environment in *.step format was imported into the LS Prepost software, moved along the x and y-axes to obtain the coordinate system's compatibility with the one shown in Fig. 2 and Fig. 4 (in the next stage of calculations). The position of the model on the Z-axis was not changed because the system was symmetrical. The units in which the model is created in Autodesk Inventor are kg and mm, hence the need for conversion to the SI system. The vehicle has been simplified to form one solid.

The vehicle model was discretized using the finite element method in the LS Prepost / LS Dyna environment. Two types of models were prepared, a solid model and a shell model to further compare the results. The solid model consists of tetrahedral elements (tetra), in the shell model triangular elements 2d were used. Because in the buoyancy problem there were not interesting stress values in the model, only interaction with the fluid/structure, the vehicle was modelled as a rigid body.

In the case of modelling solids as a rigid body with overwritten mass and moments of inertia, density, Young's modulus and Poisson's number play a secondary role. The *MAT_NULL model of the material was assumed for air and water. In table 2 materials properties are presented (in SI units).

Table 2. Air and water material parameters in SI units.

*MAT_NULL			
Air			
MID	RO	PC	MU
3	1.252	-10	1.846E-05
water			
MID	RO	PC	MU
1	1000	-1.000E+04	8.684E-04

For ALE elements with *mat_null are necessary equations of state (EOS), the simplest is linear polynomial. Equations of state for air and water are presented in table 3.

Table 3. Air and water Equation of state parameters.

*EOS_LINEAR_POLYNOMIAL				
EOSID	C4	C5	E0	V0
air 2	0.4	0.4	2.533E+05	1.0
EOSID	C0	C1	E0	V0
water 4	1.013E+05	2.250E+09	0.0	0.0

The model created in LS Prepost /LS Dyna consists of 147928 shell elements, 74060 nodes and 29 discrete elements - 15 springs and 14 dampers. These elements are described using the Lagrange method used in classical computer stereomechanical and dynamic mechanical analyses. Stiffness of springs was taken from the literature [2]. 3400000 solid elements forming air and water are described by the ALE method, which allows the transfer of pressures between elements of fluid mechanics and solid state mechanics. Interaction between Lagrange and ALE elements was implemented by *Constrained_lagrange_in_solid, friction between water and cuboid was set to 0. The volume of the vehicle was modelled as a surface rigid body (the distance between any points does not change) for which properties of mass and mass moments of inertia were determined by * PART_INERTIA. Fig. 4 presents the discrete model of the vehicle with given moments of inertia.

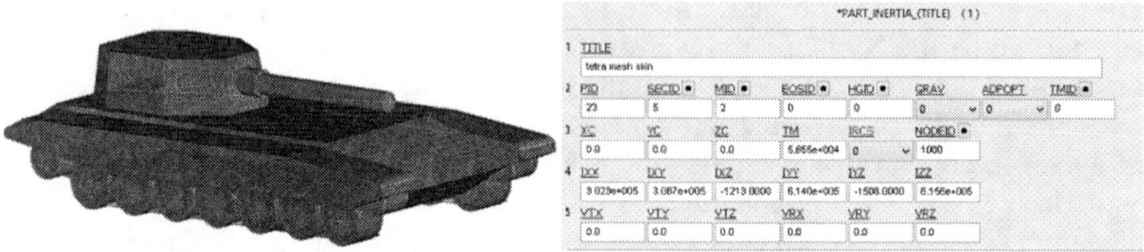

Fig. 4. The model of the vehicle along with the assigned centre of mass and moments of inertia

In node no. 1000 located in the middle of the vehicle mass, the body of the rigid moment of inertia in units of kg·m^2 and mass 56550kg, was assigned. Below is a view of Spring-specific discrete elements with a defined stiffness and dampers with a given damping coefficient. Thanks to this the vehicle could freely cross the obstacle on numerical simulations. A rigid body with given displacements along axes *X*, *Y* and *Z* connected to the vehicle through a discrete element spring with higher stiffness simulates a winch with a velocity equal to 2 m / s in accordance with design requirements.

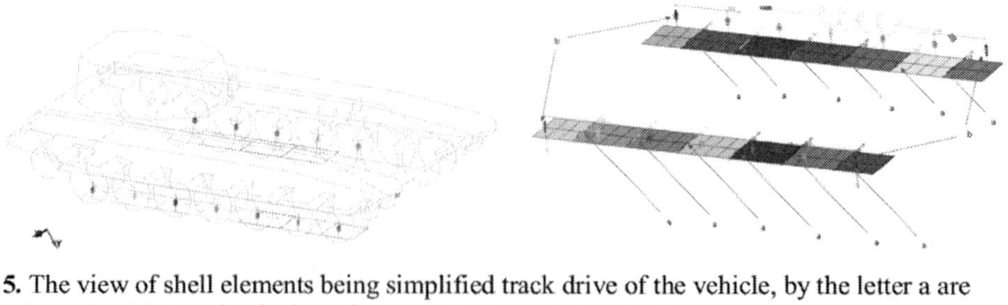

Fig. 5. The view of shell elements being simplified track drive of the vehicle, by the letter a are marked rotational joints, by the letter b are marked translational joints.

4 Tracked vehicle crossing dry obstacle

In crossing dry obstacle there is no need to have ALE elements, and the simulations were very fast. Below (Fig. 6) is presented model with dry obstacle and resultant contact force between track and ground.

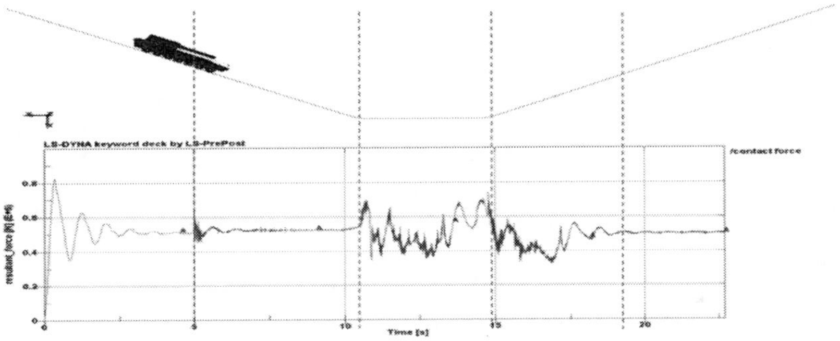

Fig. 6. Resultant total contact force distribution in a function of time, dry obstacle.

It is noticeable that the value of the contact force will stabilize before 5 s., after which the small fluctuation takes place when the object starts to move, and the total contact force is again stabilized up to 10.5 second. After this time, the vehicle descends from a plane inclined at an angle of 30 degrees to the level and goes to the straight. At this moment, the resultant value of the contact force for a short while increases and after this is decreasing to the lowest value. This takes place despite the lack of influence of hydrostatic pressure on the water. After 15.1 s the vehicle goes uphill on an inclined plane inclined at an angle of 30 degrees to the level, which gives value fluctuations and stabilization after 19 s.

5 Tracked vehicle crossing water obstacle

It is noticeable that the value of the contact force will stabilize before 5 seconds, after which the object starts to move, and the small fluctuation of the value is visible. In 10 s. is noticeable stabilization. The value of the resultant force is smaller because the buoyancy force appears. After this time, the vehicle descends from a plane inclined at an angle of 30 degrees to the level and goes to the straight. At this point, the resultant value of the contact force decreases to the minimum value. The influence of water buoyancy is visible. Before 15 seconds, the vehicle enters a plane inclined at an angle of 30 degrees to the level, which gives value fluctuations and stabilization after 19 seconds. The value of the contact force distribution depending on time was shown in Fig. 7.

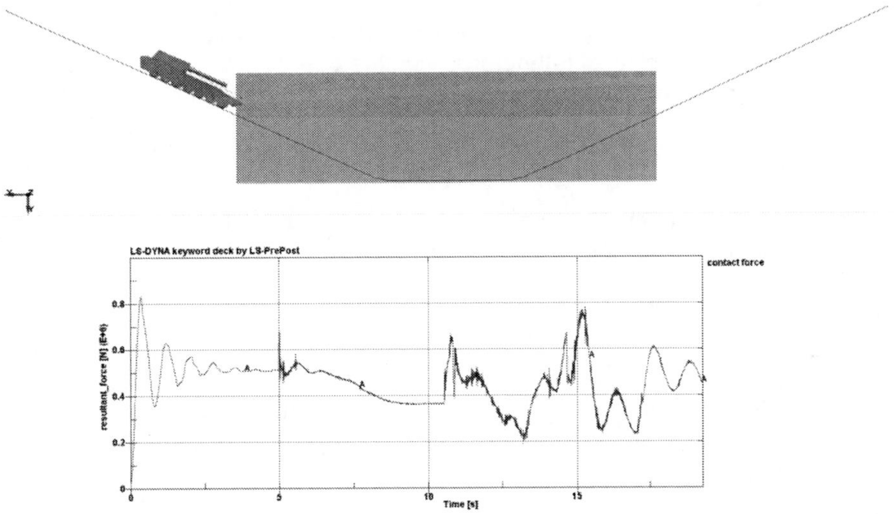

Fig. 7. Resultant total contact force distribution in a function of time, water obstacle.

6 Conclusion

LS DYNA solver was chosen for analysis of the water influence on the tracked vehicle during passing through water obstacle. The body of the vehicle was treated as a rigid body mounted by springs and dampers on tracks, and the Vehicle model was created with by shell elements with assigned moments of inertia and mass in the node, located in the centre of gravity of the vehicle. The characteristics of the resultant contact force of vehicle travel on the rigid ground for a dry obstacle and an obstacle water filled was presented and the influence of buoyancy was noticed.

REFERENCES

1. M. Souli, A. Ouahsine, L. Lewin. *ALE formulation for fluid-structure interaction problems.* Computer methods in applied mechanics and engineering, 190, 659-675, (2000)
2. J. Gniłka, A. Mężyk. Experimental identification and selection of dynamic properties of a high-speed tracked vehicle suspension system. Exploatacja i Niezawodność – Maintenance and Reliability, **19** (1): 108-113, (2017)
3. L. Olovsson, M. Souli, I. Do. LS-DYNA -ALE capabilities (Arbitrary-Lagrangian-Eulerian) Fluid-Structure Interaction Modeling, Livermore Software Technology Corporation, (2003)
4. Guenter, H. Hohl, Military terrain vehicles, Journal of Terramechanics, Vol. **44**, Is. 1, (2007)
5. Helvacioglu, S., et al., Improving the river crossing capability of an amphibious vehicle, Ocean Engineering, Vol. **38**, Is. 17–18, 2011.
6. Krueger B., Wirtz S., Scherer V., Measurement of drag coefficients of non-spherical particles with a camera-based method, Powder Technology, vol **278**, 157-170, (2015)
7. Xianzhi S., Zhengming X., Gensheng L., Zhaoyu P., Zhaopeng Zhu, A new model for predicting drag coefficient and settling velocity of spherical and non-spherical particle in Newtonian fluid, Powder Technology 321, 242-250, (2017)
8. Nakisa M., Maimun A., Ahmed Y.M., Behrouzi F., Tarmizi A., Numerical estimationof shallow water effect on multipurpose amphibious vehicle resistance, Journal of Naval Architecture and Marine Engineering, **14**, 1-8, (2017)
9. Vassilakos G.J., Elemental Water Impact Test: Phase 1 20-inch Hemisphere, NASA/CR-2015-218679
10. Kotowski M., Barnat W., Experimental and numerical buouyancy analysis of tracked military vehicle, Journal of Kones Powertrain and Transoprt, Vol. **19**, No. 4 (2012)
11. Chengzhong H., Wei B., Study on water entry problem using a level-set immersed boundary method, 5th International Symposium on Physics of Fluids, International Journal of Modern Physics: Conference Series, Vol. **34**, 1460376-11 pages, (2014)
12. J. Mohd-Yusof, Combined immersed boundary/B-spline method for simulations of flows in complex geometries, in Center Annual Research Briefs, NASA Ames/Stanford University, p. 317 (1997)

MATEC Web of Conferences **285**, 00015 (2019) https://doi.org/10.1051/matecconf/201928500015
DYNKON 2019

Damping parameters of a steel structure with a viscoelastic layer in the base node

Zdzisław Pawlak[1,*], and *Dawid* Zwoliński[1]

[1] Faculty of Civil and Environmental Engineering, Poznań University of Technology, Piotrowo 5 Street, 60-965 Poznań, Poland

Abstract. The aim of the article is to estimate the damping properties of a steel structural element equipped with a viscoelastic layer (VS) in the base node. The viscoelastic material has a good energy dissipation properties, has a relatively high strength, and its properties remain unchanged over the time. However, its damping properties may vary for different temperatures and excitation frequencies. In this work, the effect of frequency was examined, but the influence of temperature was omitted. For the selected viscoelastic material, laboratory tests were carried out to estimate its ability to dissipate energy in various load states. The viscoelastic material was then used as a washer in the joint connecting the steel element to the concrete base. The dynamic test and numerical analysis carried out for the structure with the viscoelastic layer allow estimation of some damping parameters for the considered system.

1 Introduction

In civil engineering, various methods are used to suppress excessive vibrations caused by dynamic loads. In building structures exposed to wind or earthquakes, passive and active damping systems are commonly used. Passive control techniques are often applied because they do not require any external energy supply for proper operation. Different kinds of mechanical devices, such as viscous dampers, viscoelastic dampers (VE), tuned mass dampers or base isolation systems, are applied in buildings or structures as a passive damping system. In the case of viscoelastic damper, usually a special devices containing a viscoelastic material are introduced into the bracing system of the building [1,2]. Among the others, there is a wide family of so-called base isolation systems, where the flexible isolation devices are placed horizontally at the base of the structure to decouple it from the ground [3] and to increase the natural period of vibration.

The elastomer layers are widely used as the isolation system of the base of machines producing noise or vibration. In a similar way, they are used to protect the structural elements of the building as elastomer bearings. In reinforced concrete structures (RC), elastomers are used in supports to prevent transmission of vibrations or to allow the element to deflect. When the viscoelastic material is used in a steel construction, the node is formed in such a way that the energy dissipation is obtained by the shear deformation of

* Corresponding author: zdzislaw.pawlak@put.poznan.pl

the elastomer [4]. In this article, the viscoelastic material is used as a layer inserted under the steel plate of a structural node. First, the polymer sample was tested in a testing machine, followed by a dynamic test for a steel structure with a polymer layer in the node.

2 Laboratory test

The basic properties of the polymer material were determined using the ElectroPuls™ E10000 dynamic test instrument (Instron). The dynamic analysis of the structure equipped with a viscoelastic layer was carried out using the Sirius data acquisition system.

2.1 Polymer properties

For the laboratory tests, polyurethane with the trade name B-85 was chosen as a viscoelastic material. Some mechanical properties of the material guaranteed by the manufacturer are shown in Table 1.

Table 1. Mechanical properties of B-85 polyurethane.

Characteristic	Value
Hardness [ShA]	89
Tensile strength [MPa]	31
Compression set (70°C/24h) [%]	15
Compression set (20°C/72h) [%]	14
Rebound [%]	43

The above-mentioned properties specified by the manufacturer do not determine the ability of the polymer to dissipate energy. For this reason, the sample material was subjected to an additional laboratory test.

2.2 Hysteresis loops

A dynamic test of the initially compressed sample was carried out. A round plate with a diameter of 60 mm and a thickness of 12 mm was compressed with a force of different values: 0.5, 1.0, 2.0 and 4.0 kN. Then the compression changes in time according to the sine function, the force with the amplitude of 500 N operates with different frequencies: 1.0, 5.0, 10.0 and 20.0 Hz. Thus, for the controlled input force, the magnitude of the displacement and the phase angle were measured as a response. The ElectroPuls™ E10000 test instrument must be tuned to the properties of the sample, during this process the stiffness factor is measured, in this case it was $k_z = 13.187\ \text{kN/mm}$. Taking into account the dimensions of the sample, it was possible to determine the Young's modulus of the tested material, which was $E = 56.0$ MPa. The results of the experiments are presented in the appropriate graphs in Figures 1 and 2. Figure 1 shows exemplary hysteresis loops for a compression force of 1.0 kN and for several selected frequencies.

The diagrams presented in Fig. 1 show only the change in the compression force value in the range of its amplitude. When the initial value was -1.0 kN and the amplitude was equal to 500 N, the compression varied from -0.5 to -1.5 kN. It can be observed that as the frequency increases, the displacement amplitude decreases slightly, but the graphs are very similar for the tested range of 1 to 20 Hz.

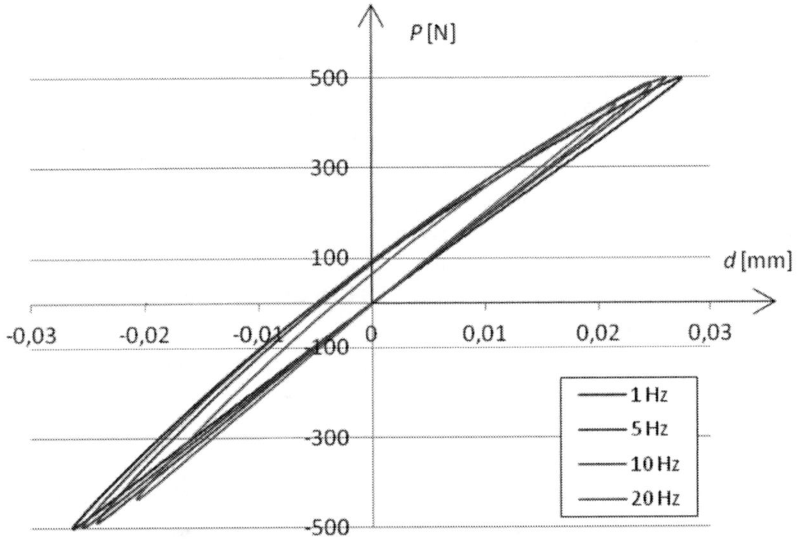

Fig. 1. Hysteresis loops for a polymer compressed by 1.0 kN force.

The next graph (Fig. 2) presents the hysteresis loops for different compression forces. To compare the loops for different forces, the diagram only shows the displacement increment in relation to the force increment.

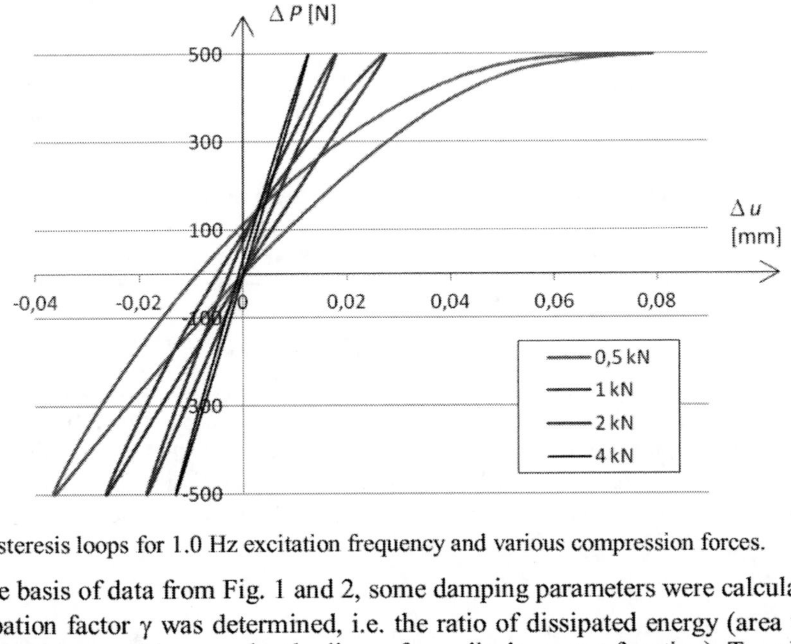

Fig. 2. Hysteresis loops for 1.0 Hz excitation frequency and various compression forces.

On the basis of data from Fig. 1 and 2, some damping parameters were calculated. First, the dissipation factor γ was determined, i.e. the ratio of dissipated energy (area within the loop) to elastic energy (area under the linear force-displacement function). To calculate the damping parameter, the hysteresis loops must be formed for the actual, measured compression force, i.e. starting from 0.5, 1.0, 2.0 and 4.0 kN, respectively. The results for selected frequencies and forces are presented in Table 2.

Table 2. Damping parameters for selected compressions and excitation frequencies.

Force *N* [kN]	0.5	1.0				2.0	4.0
Compression σ [MPa]	0.1768	0.3537				0.7074	1.4147
Frequency *f* [Hz]	1.0	1.0	5.0	10.0	20.0	1.0	1.0
Dissipation factor γ [%]	30.86	7.64	7.82	8.22	3.37	7.73	3.89
Damping ratio ζ [-]	0.0246	0.0061	0.0062	0.0065	0.0027	0.0062	0.0031

In the case of a model with viscous damping, the damping ratio $\zeta = c/c_{cr}$ is usually determined, where c_{cr} is the critical damping. There is a relationship between the damping ratio ζ and the dissipation factor γ [5]:

$$\gamma = 4\pi\zeta\sqrt{1-\zeta^2} \tag{1}$$

For a relatively small damping properties, the damping ratio is proportional to the dissipation factor, i.e. an approximate relationship is satisfied: $\zeta = \gamma/4\pi$. The estimated values of the damping ratio γ for the tested material are shown in Table 2. In addition, for a sample with an area of 28.74 cm^2, compressive forces induce appropriate stresses. It can be concluded that the damping parameters decrease with the increasing of compression.

3 Analytical and numerical approach

In the analytical approach, the reference structure was analyzed, i.e. a simple column fixed at the bottom. Natural frequencies have been derived assuming that the column made of IPE 140 profile with a height of 1.04 m is rigidly attached to the base, and a mass of 33.2 kg is at its top. The frequencies and modes of vibration were determined analytically, adopting the cantilever beam model, as well as numerically using two programs: ABAQUS/CAE and Robot™ Structural Analysis (Table 3).

Table 3. Natural frequencies for a rigidly supported column.

Frequency	**Analytical approach**	**Robot**	**ABAQUS**
f_y [Hz]	12.76	13.07	13.19
f_x [Hz]	44.27	45.36	43.90

In the numerical approach, the spatial model was tested to determine modes of vibration (see Fig. 3) and to properly compare the results obtained with different methods. Figure 3 shows the modes obtained in the ABAQUS program. In addition to the modes along the *x* and *y* axes, there is another mode, torsional, related to the frequency whose value is between frequencies f_x and f_y (bending modes).

4 Damping parameters

In laboratory tests, the damping parameters of the structure were determined by examining the signal collected by the Sirius data acquisition system. For this purpose, the Dewesoft software was used, which automatically determines the modal parameters: natural frequencies and damping coefficients. In addition, the measured response signal was used to determine the logarithmic damping decrement.

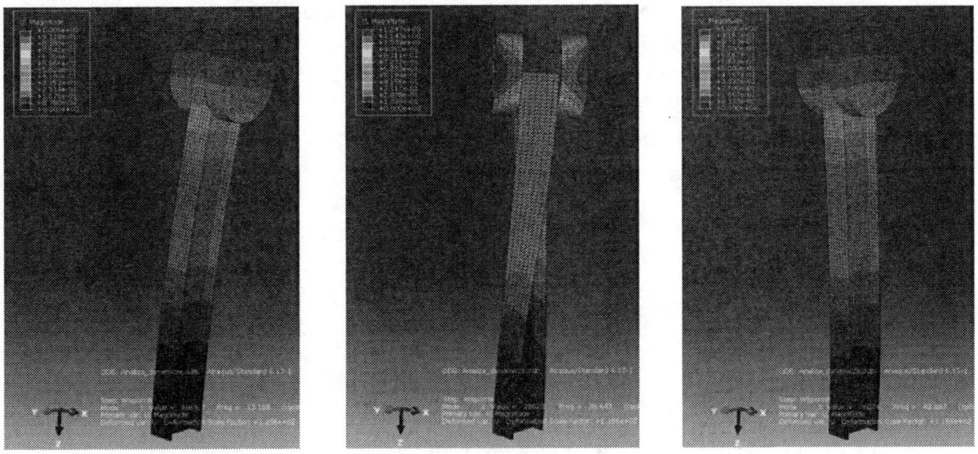

f_y=13.19 Hz f_{xy}=26.64 Hz f_x=43.90 Hz

Fig. 3. The first three modes of vibration received in the ABAQUS program.

4.1 Modal analysis

In order to determine the modal parameters of the considered structure, the Sirius data acquisition system and the Dewesoft modal analysis software were used. Modal analysis is a way to study the dynamic properties of the considered systems in the frequency domain. Dynamic properties of the system are determined for specific frequencies, i.e. natural frequencies.

4.1.1 Theoretical background

The data acquisition system Dewesoft estimates the damping properties of analysed structure taking into account the transfer function and two different models for damping. For a structural damping model it is assumed, that the equation of motion for a single degree of freedom system (SDOF) is as follows:

$$m\ddot{x}(t) + kx(t) + ihx(t) = F_0 e^{i\omega t} \tag{2}$$

where, $x(t)$ is the displacement, F_0 is the force amplitude, m stands for mass, k and h are the elements of complex modulus, which could be interpreted as a storage and a loss modulus for considered material (i.e. $k^* = k + ih = k(1 + i\eta)$), $i = \sqrt{-1}$ and ω is the natural frequency. In the experiment, the system response to the sinusoidal input given in the time domain is transferred to the frequency domain by Fourier transform. In this way, we get an output signal to a sinusoidal input of the same frequency but with different amplitude and phase. The frequency response function (FRF) is defined as the magnitude and phase difference between input and output sine waves at different frequencies. The so-called receptance of the frequency response function is defined as the ratio of the measured displacements $X(\omega)$ and force $F(\omega)$:

$$\alpha(\omega) = \frac{X(\omega)}{F(\omega)} = \frac{1}{k - \omega^2 m + ih} = \frac{1}{\omega_r^2\left[1 - \left(\frac{\omega}{\omega_r}\right)^2 + i\eta_r\right]} \tag{3}$$

where, η_r is the damping loss factor that corresponds to the resonance frequency ω_r. To plot the frequency response, we create a vector of frequencies **w** and compute the value of the transfer function $TF(i\mathbf{w})$ at those frequencies. Since the elements of transfer function are complex number, we can plot both its magnitude and phase as the Bode plots or we can plot its position in the complex plane as the Nyquist plot [6]. The receptance forms a perfect circle on the Nyquist plot, even if the multi-degrees of freedom systems are considered, because the contribution of other modes is irrelevant in the vicinity of the resonance. After fitting the circle to N points around the resonance peak, one can take two points $(\omega_b\,, \omega_a)$ from this circle, below and above the resonant frequency ω_r and calculate the modal damping (loss factor η_r) as:

$$\eta_r = \frac{\omega_a^2-\omega_b^2}{\omega_r^2[\tan(\theta_a/2)-\tan(\theta_b/2)]} = \frac{\omega_2^2-\omega_1^2}{2\omega_r^2} \approx \frac{\omega_2-\omega_1}{\omega_r} \tag{4}$$

where, ω_1 and ω_2 are the half power points of the considered FRF, i.e. $\alpha(\omega_1) = \alpha(\omega_2) = \alpha(\omega_r)/\sqrt{2}$.

The equation of motion for SDOF system with a viscous damping model is

$$m\ddot{x}(t) + c\dot{x}(t) + kx(t) = F_0\sin(\omega t) \tag{5}$$

where, c(t) stands for the viscous damping parameter of considered system. In this case, the so-called mobility of the frequency response function is used, which is understood as the ratio of the measured velocity $\dot{X}(\omega)$ and force $F(\omega)$:

$$Y(\omega) = \frac{\dot{X}(\omega)}{F(\omega)} = \frac{i\omega}{k-\omega^2 m+i\omega c} = \frac{i\omega}{\omega_r^2\left[1-\left(\frac{\omega}{\omega_r}\right)^2+2i\left(\frac{\omega}{\omega_r}\right)\varsigma_r\right]} \tag{6}$$

where, $\zeta_r = c/c_{cr}$ is the damping ratio which corresponds to the resonance frequency ω_r, while c_{cr} is the critical damping. The mobility of FRF presented on the Nuquist plot leads to a convenient circle-fitting method that allowes determining the resonance frequency and damping ratio:

$$\zeta_r = \frac{\omega_a^2-\omega_b^2}{2\omega_r\left[\omega_a\tan(\theta_a/2)+\omega_b\tan(\theta_b/2)\right]} = \frac{\omega_2^2-\omega_1^2}{4\omega_r^2} \approx \frac{\omega_2-\omega_1}{2\omega_r} \tag{7}$$

The symbols used in Eqn 7 are the same as in Eqn 4. In addition, there is a relationship between the damping loss factor and damping ratio: $\eta = 2\zeta$ [7].

4.1.2 Results of the test

The dynamic test was carried out on a steel structure, a column made of I-section (IPE 140) fixed at the bottom (see Fig. 4a) was selected. A mass of 33.2 kg was attached to the top of the column at the height of 1.04 m. The connection to the base was made using M16 anchor bolts. During the test an additional polymer washers were placed under the column endplate (Fig. 4b). The dimensions of the endplate were 20x140x250 mm, and the polymer washers were 12 mm thick.

In the dynamic test, the relationship between the compressive stress value in the polymer and the dynamic parameters of the analyzed structure was investigated. The electronic torque wrench was used to control the torque as a bolt tension indicator. In this way, different values of the compressive stresses were obtained in polymer washers.

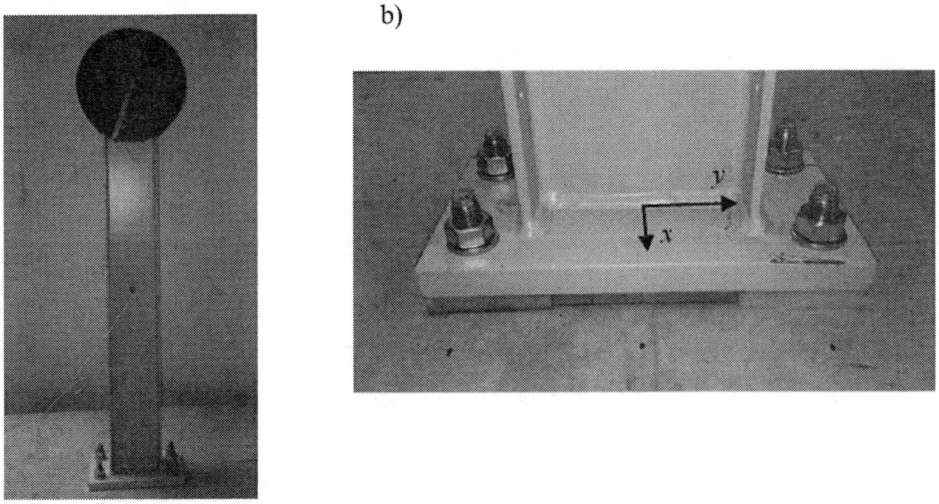

Fig. 4. Steel column with attached mass: a) view of the system, b) polymer washers in the node.

In order to obtain a similar compression stress in the polymer as in the previous test (Table 2), the required torque was calculated (Table 4).

Table 4. Relationship between torque, tension in the bolt and stress in the polymer.

Torque, *Mo* [Nm]	4.92	9.85	19.69	39.39	78.77
Bolt tension, *F* [kN]	1.733	3.466	6.932	13.864	27.728
Compression, σ [MPa]	0.1768	0.3537	0.7074	1.4147	2.8294

The torque value was calculated on the basis of the guidelines for M16 pretension bolts, class 8.8. The relationship between the preload and the torque was adopted for a coefficient of friction of 0.18. In the dynamic test, the tension of the bolts was changed according to the data in Table 4, and thus an adequate compression in the polymer washers was obtained. The column was excited in two perpendicular planes with a modal hammer and the response of the structure in the form of acceleration was measured. Using the Dewesoft software and the modal identification function (MIF), the natural frequencies ω_r were determined. Then, using the receptance function (Eqn. 3) and the mobility function (Eqn. 6), the appropriate values of the loss factor η_r and damping ratio ζ_r were determined (Table 5).

Table 5. Damping coefficients of the first mode for different compressive stresses in the polymer.

Compression σ [MPa]	**Torque *Mo* [Nm]**	**Model of damping**	f_y **[Hz]**	η_y **[-]**	ζ_y **[-]**
0.1768	4.29	structural	9.592	0.2263	
		viscous	9.257		0.1150
0.3537	9.85	structural	10.213	0.1999	
		viscous	10.314		0.1020
0.7074	19.69	structural	10.463	0.1131	
		viscous	10.460		0.0625
1.4147	39.39	structural	10.962	0.0850	
		viscous	10.815		0.0425

When a receptance function was used, a structural damping model was assumed, for the mobility function a viscous damping model was applied. It can be seen that the damping coefficients approximately meet the condition $\eta = 2\zeta$.

4.2 Logarithmic decrement of damping

The dynamic characteristics of the structure were also determined directly from the measured response signal, i.e. the acceleration function in the time domain. The use of a filter that narrows the signal to a resonant frequency leads to a sine wave function whose amplitude decreases over time. Taking the values of two successive amplitudes $q(t)$ and $q(t+T)$, one can calculate the logarithmic damping decrement:

$$\delta_r = \ln\left(\frac{q(t)}{q(t+T_r)}\right) = \frac{2\pi\zeta_r}{\sqrt{1-\zeta_r^2}} \tag{8}$$

After calculating the damping decrement δ_y, the value of the damping ratio ζ_y from the Eqn. 8 can be determined. The results for the first vibration mode and selected compressions in the base (*Mo* torque), with and without polymer, are shown in Figure 5.

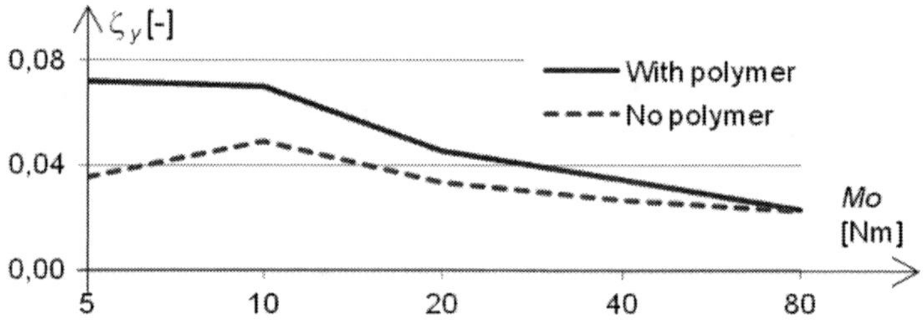

Fig. 5. The first mode damping ratio as a function of torque, causing force in the anchor bolt.

5 Conclusions

The paper presents several approaches and techniques for describing damping in structural systems. Damping coefficients: dissipation factor, loss factor, damping ratio and logarithmic decrement were estimated on the basis of data taken from laboratory measurements. The use of different calculation methods leads to slightly different results.

Moreover, the influence of polymer washers used in the butt nodes on the dynamic characteristics of considered structure was investigated. As expected, the increase in compressive stress in the polymer washers reduced their damping properties. But even with relatively high compression, the considered polymer has the ability to dissipate energy.

References

1. K. C. Chang, T. T. Soong, S. T. Oh, M. L. Lai, JSE, 121, 10, 1418-1425 (1995)
2. M. P. Singh, T. S. Chang, J. Eng. Mech., 135, 571-580 (2009)
3. G. P. Warn, K. L. Ryan, Buildings, **2**, 300-325 (2012)
4. A. Banisheikholeslami, F. Behnamfar, M. Ghandil, JCSR 117, 185-195 (2016)
5. J. N. Libii, WIETE, **7**, no1, 28-33 (2009)
6. W. S. Levine, *The control handbook* (CRC PRESS, IEEE PRESS, 1996)
7. FRF-modal test manual. www.dewesoft.com

Dynamic performance of a vehicular bridge with lightweight FRP composite structural elements

Tomasz Siwowski[1,*], *Mateusz* Rajchel[1]

[1]Rzeszow University of Technology, Department of Roads & Bridges, 35-395 Rzeszow, Poland

Abstract. The lightweight nature and low stiffness of fibre reinforced polymer (FRP) structures combined with low material damping may lead to excessive structural vibration and may increase the dynamic amplification of deformations and stresses in FRP bridge structures. The paper provides a short review on dynamic characterization of existing vehicular bridge superstructures made of FRP structural components. Against the backdrop of a worldwide review the paper presents a field study on the dynamic characteristics of the hybrid FRP – concrete bridge superstructure recently built in Poland. The dynamic amplification factor, the first natural frequency and the damping ratio were experimentally determined in static and dynamic load tests and compared with the relevant codes and requirements. The paper concludes by summarizing challenges related to structural dynamics of bridge superstructures with FRP structural components.

1 Introduction

Since the beginning of the XXI century the fibre reinforced polymer (FRP) composites have become an integral part of the construction industry because of their versatility, high strength-to-weight ratio, enhanced durability, resistance to fatigue and corrosion, accelerated construction as well as lower maintenance and life-cycle costs. Due to these excellent performance characteristics FRP composites offer very promising options for a wide range of civil infrastructure applications, particularly in bridge engineering [1-3].

However, relative low mass, low damping and low stiffness make FRP composite bridges sensitive to dynamic excitation, which may lead to discomfort of human occupants and larger dynamic amplification of stress and deformation than is encountered in structures made of conventional structural materials. Consequently, design might be governed by a vibration serviceability state. Lack of data on vibration performance of FRP structures and non-existence of a state-of-the-art vibration serviceability design guideline means that current practice is conservative, often meaning only short-span FRP bridge solutions are executed. To fully exploit the benefits of using FRP composites and to extend their use beyond current practice requires a better understanding of dynamic behaviour.

The paper provides a short review on dynamic characterization of constructed vehicular bridge superstructures made of FRP structural components as well as the basic

* Corresponding author: siwowski@prz.edu.pl

characteristics of a novel bridge superstructure with FRP components recently built in Poland. Against the backdrop of a worldwide review the paper presents a field study on the dynamic characteristics of the hybrid FRP – concrete bridge superstructure constructed with FRP box girders and concrete deck slab. The dynamic amplification factor (DAF), the first natural frequency and the damping ratio were experimentally determined in the static and dynamic load tests and compared with the relevant codes and requirements. The paper concludes by summarizing challenges related to structural dynamics of bridge superstructures with FRP structural components.

2 Review on dynamic characteristics of FRP vehicular bridges

Among all constructed vehicular bridges using FRP structural elements (normally FRP girders or decks), the dynamic analysis and characterization has been conducted on several cases from published data. A summary of the available information is presented in Table 1, where the dynamic amplification factors, the first natural frequencies and the damping ratios are provided. However, because not all dynamic parameters were determined in the analysed field tests, the considerable lack of data was revealed in this review.

Table 1. Dynamic characteristics of the constructed FRP vehicular bridges.

Bridge name	Structural system	Main dimensions L x B [m]	Dynamic characteristics			Source
			DAF	first natural frequency [Hz]	damping ratio [%]	
Tom's Creek Bridge	FRP girders, wooden deck	5.33 x 6.70	1.90	N/A	N/A	[4]
Route 601 Bridge	FRP girders, glulam deck	11.90 x 8.54	1.36	N/A	N/A	[5]
Laurel Lick Bridge	FRP girders, FRP deck	5.79 x 4.88	1.93	N/A	N/A	[6]
Bennetts' Creek Bridge	FRP slab	7.81 x 10.14	1.52	26.8	4.53	[7]
Nowa Wies Bridge	FRP girders, FRP deck	10.00 x 7.70	1.10	10.1	4.5	[8]
Toowoomba Bridge	Hybrid FRP-concrete beams	10.0 x 5.0	N/A	9.5	2.0	[9]
Tides Mill Stream Bridge	Hybrid FRP-concrete beams	13.50 x 9.90	1.23	N/A	N/A	[10]

It is clearly seen from Table 1 that DAF of FRP bridges ranges from 1.10 to 1.93. The majority of the obtained DAFs exceeded the design values as shown in the relevant national standards (typically in the range of 1.25 – 1.33). There has been no consensus made on how to implement the DAF in designing FRP bridge superstructures. The first natural frequency ranges from 9.5 to 26.8 and is much higher (means better) than recommended values according to the relevant standards, typically above 3 Hz. However, the damping ratio in the range of 2.0 – 4.53 % is considerably smaller than the corresponding values for concrete bridges, which are typically about 8 %.

The experimental results of the studies shown in Table 1 demonstrate that FRP bridge dynamic behaviour has not been completely recognized. Due to small number of existing results and large discrepancies between them, the dynamic characteristics of bridge superstructures consisting of lightweight FRP structural elements should be still investigated and analysed. Moreover, only a very limited number of research studies have been conducted on FRP-concrete hybrid bridges. In addition, a global structural integrity and health condition evaluation is necessary to ensure the long term performance of the key structures made of FRP composites. Evaluating integrity and health condition of FRP bridges through structural dynamic characterization could be a promising technique to be developed. Therefore the acceptance of such FRP bridge designs requires much more field evaluation of their structural performance. The comprehensive testing of the actual FRP – concrete superstructure under dynamic loading provides much improved insight and understanding of this unique structural system.

3 Description of the bridge superstructure

The hybrid FRP composite – concrete vehicular bridge is situated in southeast part of Poland along a local road [11]. This is a 22.0 m long single-span simple supported bridge with 10.5 m wide deck carrying 2 × 3.5 m wide roadway and 1 × 2.0 m wide sidewalk. Its nominal carrying capacity is 40 tonnes according to the Polish bridge standard. The bridge superstructure is formed by four FRP composite box girders with 0.18 m thick concrete deck slab acting compositely with the girders (Fig. 1). The deck equipment consists of two concrete sidewalk slabs with safety barriers, polymer curbs, conventional insulation and stone mastic asphalt (SMA) pavement layer, drainage and expansion joints.

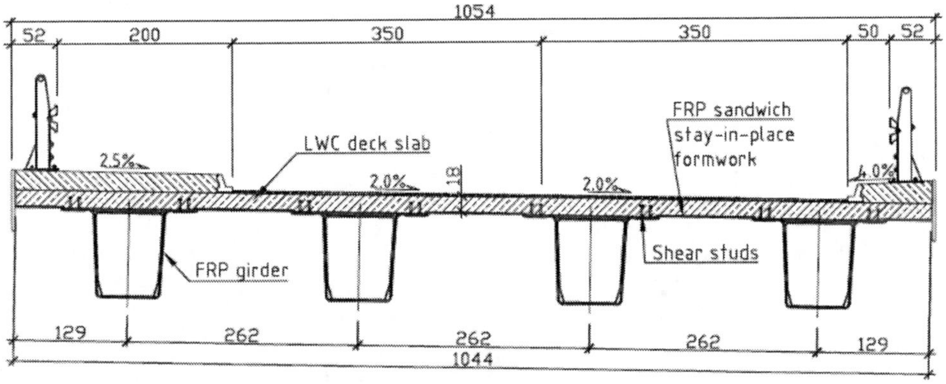

Fig. 1. Cross-section of the FRP composite - concrete bridge superstructure.

Each girder consists of two main parts: a FRP composite box with trapezoid open cross-section and a concrete slab. Steel shear connectors are used to ensure a composite action between FRP composite and concrete. The FRP composite box is 1025 mm deep, 1550 mm wide at the top and 621 mm wide at the bottom. The both top flanges of the shell are 301 mm wide and are connected with the bottom plate by means of two slightly inclined webs. The individual laminates have the thickness of about 25 mm, 22 mm and 19 mm for top flanges, webs and bottom plate, respectively. The top flanges and webs are made of solid and sandwich GFRP respectively, whereas the bottom plate has a hybrid GFRP/CFRP solid structure. The sandwich webs consist of 15 mm thick polyvinyl chloride (PVC) foam core and two 3.5 mm thick GFRP face laminates. To increase the torsional stiffness of the FRP box and to prevent shear-bending buckling of its webs, internal diaphragms are placed along the length of the girder. The diaphragms are made of sandwich panels with the structure similar to webs.

The 180 mm thick concrete deck slab is made of 35/38 class lightweight concrete (LWC) based on *lytag* type aggregate and reinforced longitudinally and transversally with two grids of 12 mm GFRP bars. The LWC slab is connected to FRP box girders by means of shear connectors made of galvanized steel bolts welded to small steel plates and fastened to box top flanges with epoxy adhesive. The 950 mm long support zones of each girder are filled with LWC to form support end cross-beams and to ensure transverse stiffness of entire superstructure. Steel plates with welded steel bolts are fastened to the webs inside each FRP box to ensure a composite action between FRP composite and concrete in the support zones.

Finite element (FE) structural analysis, design according to new European recommendations as well as construction process of the bridge have been briefly described elsewhere [11, 12].

4 Dynamic testing of the bridge

Three dynamic tests were performed to measure dynamic characteristics of the superstructure by driving the truck with an approximate total load of 320 kN along the bridge with a speed of 10, 30 and 50 km/h (schemes D1 to D3, respectively). To increase the dynamic effect two additional tests were also performed: test with speed of 30 km/h and sudden braking on the deck in midspan (scheme D4) and test with a speed of 10 km/h and the 0.05 m deep obstacle placed transversally on the deck in midspan (scheme D5).

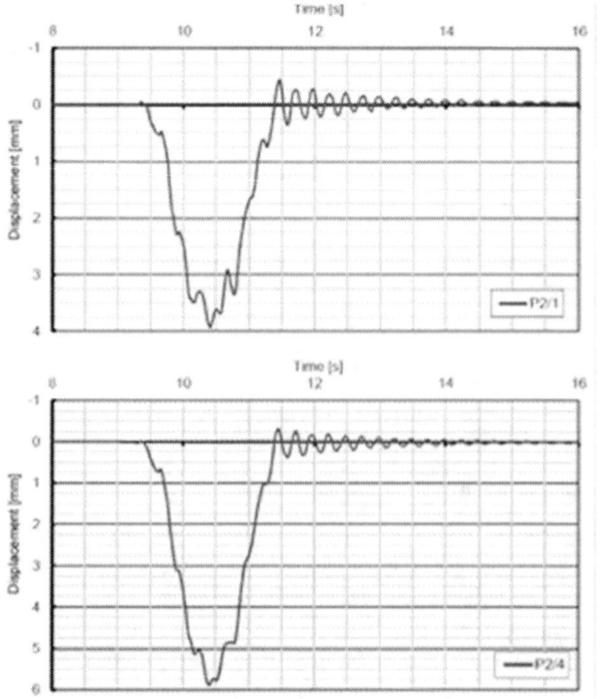

Fig. 2. Dynamic response in the midspan of both external girders under D3 load scheme (50 km/h).

During dynamic load tests the displacements and accelerations of the hybrid superstructure were measured at six points: four linear variable differential transducers

(LVDTs) were mounted to girders in the midspan (measurement points P2/1 to P2/4 for all four girders, respectively) and two piezoelectric accelerometers were attached to deck slab in the midspan. The digital data acquisition system HBM Spider with Quantum X amplifier was used to record all relevant measurements. Dynamic responses in the midspan of both external girders under D3, D4 and D5 load schemes are shown in Figs 2, 3 and 4, respectively.

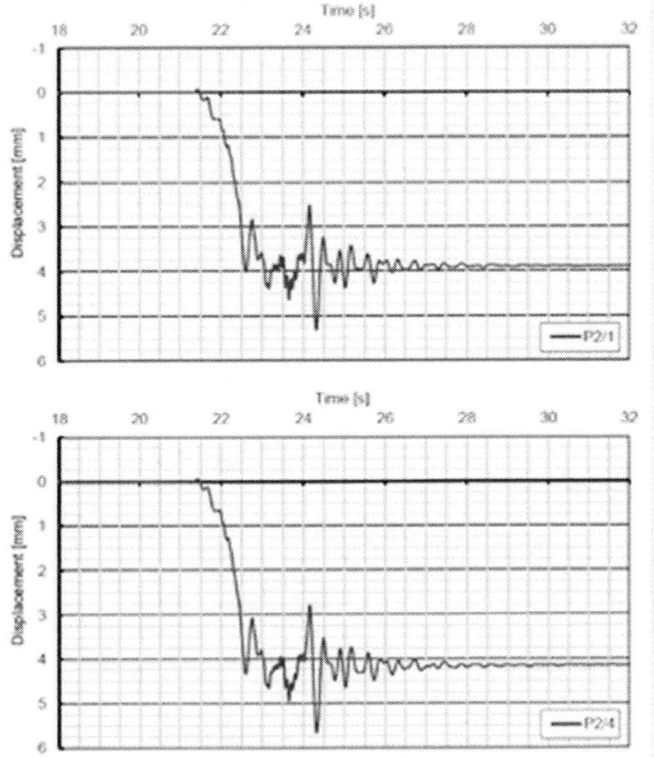

Fig. 3. Dynamic response in the midspan of both external girders under D4 load scheme (30 km/h + braking).

The dynamic amplification factor accounts for the amplification of the design static live load due to a moving vehicle crossing a bridge. This factor is typically described as a function of the interaction of the bridge and vehicle and is influenced by properties such as mass, stiffness, vehicle speed, and surface roughness. The DAF for each load scheme was calculated as the ratio of maximum dynamic displacement (read directly from the "time - displacement" plots) and maximum static displacement, measured under the relevant truck positioned in the midspan (static schemes, not reported here). Basing on measured displacements, the DAFs for all vehicle speeds (10/30/50 km/h) were assessed as 1.055/1.104/1.063, respectively. All these values are much lower than the DAF used in bridge design (1.25). Moreover, the determined factors are much lower than the average dynamic amplification factor of the similar US hybrid bridge (see Table 1 and [10]) equalled as much as 1.23 in the most heavily loaded girders.

The first natural frequency of the hybrid superstructure was determined by means of the "time-displacement" plot for scheme D5 with the transverse obstacle on the deck slab (Fig. 4) as the ratio of number of displacement peaks in time t_s and this time t_s (in seconds). The estimated first natural frequency of the hybrid superstructure equalled 3.98 Hz and was higher (means better) than recommended value according to the performance requirements for road bridges in Poland (minimum value 3.0 Hz). Finally, the experimentally determined logarithmic decrement δ was determined by means of the "time-displacement" plot for scheme D4 with braking (Fig. 3). The obtained value δ=0.201 was used for estimating span's damping ratio, which relevant value amounts 3.2%. As might be expected this value is much lower than corresponding damping characteristics for a typical concrete bridge with similar beam–slab superstructure and span length.

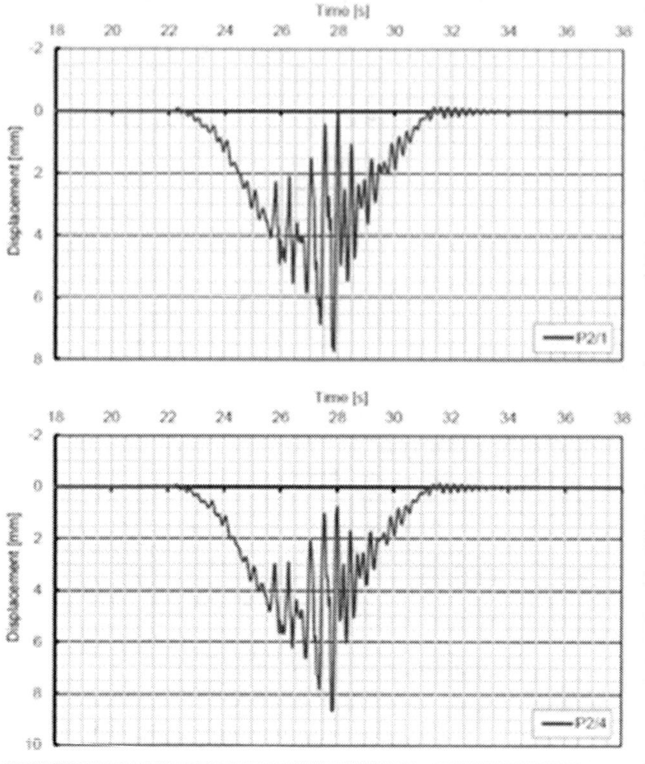

Fig. 4. Dynamic response in the midspan of both external girders under D5 load scheme (10 km/h + transverse obstacle).

5 Summary and conclusions

This investigation focused on evaluating the dynamic performance under proof load of a newly constructed hybrid FRP composite – concrete bridge superstructure. The bridge superstructure was instrumented with a series of LVDTs and accelerometers to measure the relevant dynamic characteristics. The dynamic tests were conducted using single truck weighing 320 kN and moving with various speeds. The results obtained from the experimental investigation were used to determine three key dynamic performance characteristics: dynamic amplification factor, first natural frequency and damping ratio.

The main findings of the study are summarized as follows: DAFs for speeds 10/30/50 km/h were assessed as 1.055/1.104/1.063, respectively; estimated first natural frequency equalled 3.98 Hz and experimentally determined logarithmic decrement δ=0.201 enabled to estimate the span's damping ratio as 3.2%. All these parameters revealed very good dynamic behaviour of the hybrid bridge superstructure. Moreover, with respect to the dynamic performance, the novel hybrid system fully conformed to the provisions of the domestic bridge standards and the new European recommendations for FRP bridges [13].

To use FRP structures in vehicular bridge, the dynamic characteristics of the bridge superstructure should be carefully examined. The value of the dynamic coefficient has not been well established for FRP bridges. Generally, the analytical and experimental process used for conventional bridges can be applied into FRP bridges. However, attention should be paid to the fact that the isotropic constitutive material modelling of traditional bridge analysis needs to be replaced by orthotropic constitutive material modelling. The finite element (FE) model should be calibrated against the results gathered during field testing of an actual bridge superstructure under dynamic load following the well-established and proved procedure [14, 15].

Another challenge is how to improve the damping property of FRP bridge superstructures, although this problem is common in civil engineering, the brittle nature of FRP (especially for GFRP) makes it more critical. Some research has been carried out to improve the material damping properties. A promising alternative is to improve system damping. One of them is to design a hybrid FRP – concrete structure. The dynamic tests conducted to date along with the described above have clearly revealed that hybrid FRP - concrete bridges are much better with the respect of dynamic behaviour then their all-composite counterparts. These results are also encouraging for the design of hybrid bridge systems where the initial cost of the FRP material is a substantial portion of the superstructure cost.

All these challenges necessitate the study of structural dynamics of FRP bridges. A global non-destructive evaluation technique through structural dynamic parameters is an opportunity for dynamic analysis of structural characteristics, which is also useful to ensure the long term performance of FRP bridges [16].

Acknowledgments

This work was created within the framework of the project: "*ComBridge* – An innovative FRP composite road bridge". The project was implemented as part of a pilot programme entitled: Support for Research and Development Works in the Demonstrative Scale DEMONSTRATOR+ (Contract No. DEM-1-041/001) and was co-founded by the National Centre for Research and Development, Poland (NCBiR) as well as the industry partners: Mostostal Warszawa S.A. and Promost Consulting, Rzeszow.

References

1. L. C. Holloway, P. R. Head, *Advanced polymer composites and polymers in the civil infrastructure* (Elsevier Science Ltd, London, 2001)
2. M. Zoghi (Ed.), *The international handbook of FRP composites in civil engineering* (CRC Press, Taylor & Francis Group LLC, Boca Raton, 2014)
3. T. Siwowski, *FRP composite bridges. Structural shaping, analysis and testing* (Wydawnictwo Naukowe PWN, Warsaw, 2018)
4. W. D. Neely Jr, *Evaluation of in-service performance of the Tom's Creek Bridge* (Virginia Polytechnic Institute and State University, Blacksburg, VA, 2000)
5. T. E. Cousins, J. J. Lesko, *Construction of a Virginia Short-Span Bridge with the Strongwell 36-Inch Double-Web I-Beam* - Report No. VTRC 06-CR5 (Virginia Transportation Research Council, Charlottesville, VA, 2005)
6. S. Aluri, C. Jinka, H.V.S. GangaRao, Dynamic response of three fibre reinforced polymer composite bridges, J Bridge Eng **10**, 6 (2005)

7. A. J. Aref, S. Alampalli, Vibration characteristics of a fiber reinforced polymer bridge superstructure, Compos Struct **52**, 3–4 (2001)
8. T. Siwowski, M. Rajchel, D. Kaleta, Dynamic characterisation of road bridge made of fibre reinforced polymer composites, J Civ Eng EnvArch **XXXIV**, 64 (2017)
9. G. M. Van Erp, T. J. Heldt, L. Mc-Cormick, D. Carter, C. Tranberg, Development of an Innovative Fibre Composite Deck Unit Bridge, Proc. IABSE, Melbourne, 2002.
10. D. K. Harris, J. M. Civitillo, A. Gheitasi, Performance and behavior of hybrid omposite beam bridge in Virginia: live load testing, J Bridge Eng **21**, 6 (2016)
11. M. Rajchel, T. Siwowski, Research on the hybrid FRP composite – concrete bridge girder, J Civ Eng EnvArch **XXXII**, 62 (2015)
12. T. Siwowski, D. Kaleta, M. Rajchel, L. Wlasak, The first Polish road bridge made of FRP composites, Struct Eng Int **27**, 2 (2017)
13. L. Ascione, E. Gutierez, S. Dimova, A. Pinto, S. Denton, *Prospect for new guidance in the design of FRP* - Scientific and Technical Report No. 27666 (Joint Research Centre, European Union, Brussels, 2016)
14. J. Chróścielewski, M. Miśkiewicz, Ł. Pyrzowski, M. Rucka, B. Sobczyk, K. Wilde, Modal properties identification of a novel sandwich footbridge – Comparison of measured dynamic response and FEA. Com Part B: Eng **151** (2018)
15. J. Chróścielewski, T. Ferenc, T. Mikulski, M. Miśkiewicz, Ł. Pyrzowski, Numerical modelling and experimental validation of full-scale segment to support design of novel GFRP footbridge, Compos Struct **213** (2019)
16. A. Zhou, Y. Bai, T. Keller, Dynamic characteristics of bridge superstructures with FRP composite structural elements, Proc. CCC, Lyon. (2005)

Research of influence of vibration impacts of chimes on the steeples of St. Isaac's Cathedral (Saint Petersburg) after reconstruction thereof

Elena E. Smirnova[1,*], *Sergey* N. Savin[1], and *Denis* V. Larin[1]

[1]Saint Petersburg State University of Architecture and Civil Engineering, 190005 Vtoraja Krasnoarmejskaja ul. 4, St. Petersburg, Russia

Abstract. This article gives consideration to the ways of solving a problem of ensuring safety of buildings and structures in use provided their operating conditions have changed. The assessment of possible consequences of bells installation on the steeples of St. Isaac's Cathedral is given as an illustration. The measurements of vibrations with simultaneous assessment of robustness and bearing capability of the brickwork have shown that the vibration impact of chimes of heavy bells can not adversely affect the bearing structures of the bell towers. However, there is always a risk of such adverse affect on the structure of the bell tower and stability of soils at its base. It is offered, therefore, to carry out investigation before the beginning of reconstruction using the structure model, which makes it possible to give consideration to the impact of «new» loads to exclude any possible emergency situation at the objects being reconstructed.

1. Introduction

Due to reconstruction and restoration of the original guise of the landmarked buildings the investigators are facing the problems of assessing possible adverse consequences of such sorts of work in the recent years. The most problematic are the works requiring dismantling of the bearing and fencing structures built into original scheme. Kronstadt Naval Cathedral, building of the church of Transfiguration of Jesus and other significant historical buildings have become the examples of such sort of objects for Saint Petersburg. The builders and architects have succeeded to successfully solve these complex problems using the up-to-date methods of technical diagnostics and calculation by means of three-dimensional finite-element models [1–2].

Another type of works implies a revival of ensembles of bells on the steeples [3–4]. In order to assess possible adverse consequences of these actions, it is necessary to take into account not only the new load emerging in the structures not being used for direct purpose for long, but the possible consequences of vibration impact on the soils of bases of structures.

A group of researchers (including authors of this article) under the leadership of I.V. Sitnikov from the research-and-development center No.26 of the Ministry of Defense of the Russian Federation together with the specialists of Saint-Petersburg Railway Engineering and Transportation University have conducted measurements of vibrations caused by the chimes of the bell tower of St. Isaac's Cathedral (Saint Petersburg) and analyzed the possible consequences thereof within a period of 2013-2015.

The measurements of vibrations with simultaneous determination of the robustness of brickwork of the bearing structures of steeples have been performed in two stages: in November-December 2015 in the course of installation of a big ten-ton polylinear bell on the North-eastern steeple and in December 2016 after installation of 17-ton bell on the north-western steeple of St. Isaac's Cathedral.

The level of vibrations at the bearing structures of steeples, parameters of low-frequency dynamic impacts occurring in case of bells swinging, and strength characteristics of the stonework have been measured in the process of investigations.

The acquired results have been analyzed and compared with the existing requirements of standards [5–8] with a view to exceedance of actual loads.

2. Materials and Methods

The oscillations were recorded using a research measuring complex, which allows measurements of the RMS (root-mean-square error value) of vibration velocity, relative magnitude of vibration displacement, relative displacement and

* Corresponding author: esmirnovae@yandex.ru

linear displacement and indication of the measured parameters. It consists of measuring sensors of CJSC Geoakustika A1612 and A1633, and KV-12 (Fig. 1–2). The piezoelectric A1612 and A1633 single-component accelerometers are designed to convert accelerations acting along their measuring axes into proportional electrical signals; accelerometers are used as primary transducers as part of vibration measuring systems, and can also be used for measuring low level vibrations. In A1612 units, a package of unloaded bimorph disks with central fastening is used as a sensitive element; sensitive elements A1612 and A1633 are connected to preamplifiers, which form the main characteristics (AFC – amplitude-frequency characteristic, conversion factor, supply voltage, output impedance, etc.), which allows accelerometers to be adapted to the solution of a specific measurement task.

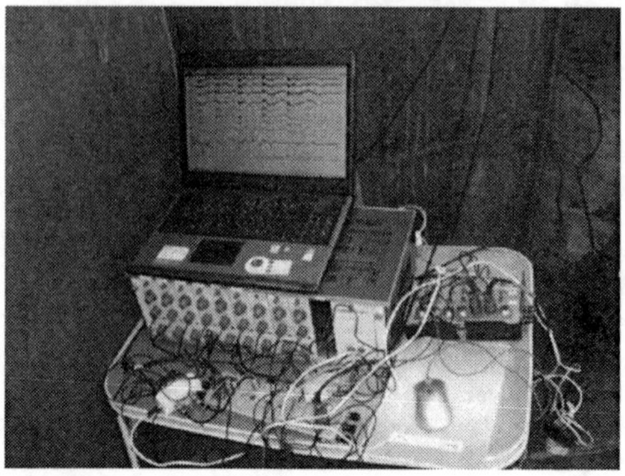

Fig. 1. Measurement Sensors A1612 and A1633 of Geoakustika CJSC (operating frequency range - from 0.2 to 400 Hz; amplification factor - up to 100 mV / m / s^2).

Fig. 2. Measurement Sensor KV-12 of Geoakustika CJSC (sensitivity - 300 mv/ms^2; frequency range - up to 400 Hz).

The broad-band amplifier Robotron is also an integral part of the system. It has the following measurement modes: acceleration, speed, displacement; its amplification ranges from 1 to 10,000 with a step change in 10 dB steps; frequency range from 1 Hz to 20 kHz; the number of measurement channels is at least 8; 220V or DC battery power supply;

- Analog-to-digital converter (ADC). Its characteristics are as follows: the number of channels is at least 8, the bit depth is at least 12 bits, the ranges of measured signals are ± 5.12 V, 2.56 V, 1.024 V; maximum conversion frequency – 300 kHz / channel.
- Measurement cables;
- Portable computer is equipped with software for input and processing of measurement information.

The means of excitation of oscillations (before installation of 17-ton bell) – external impulse (special strike). Such a blow is applied in the upper part of the object, for example, with a load of up to 50 kg.

The means of excitation of oscillations (after installation) – bell ringing.

The methodology of determining level of vibrations of building structures includes the following operations:

- registration of oscillations;
- calculation of Fourier spectra thereof;
- analysis of Fourier spectra with the aim of identifying resonance peaks corresponding to the level of vibrations and oscillations in a wide range of frequencies.

3. Results

In case of necessity of assessing consequences of vibration load impact on the bearing structures the dynamic parameters thereof have been determined in the way of:

- obtaining pulsed implementations of identified resonance peaks for every free vibration mode by means of inverse Fourier transformation;
- identification and graphical representation of various waveforms.

The vibrations have been recorded «under load» in the process of chiming.

The dynamic parameters of the «beams-bearing structures of bell tower» system have been determined by analyzing the structures response to the pulsed load. The obtained frequencies and waveforms have been analyzed, response spectra determined and values of accelerations and velocities in the targeted range of frequencies (up to 50 Hz). The photo and scheme of measurements at the structures of the North-western bell tower, where the vibrations have been registered and the brickwork probed, are shown in Fig. 3 and 4:

Fig. 3. South-eastern steeple with the installed bells.

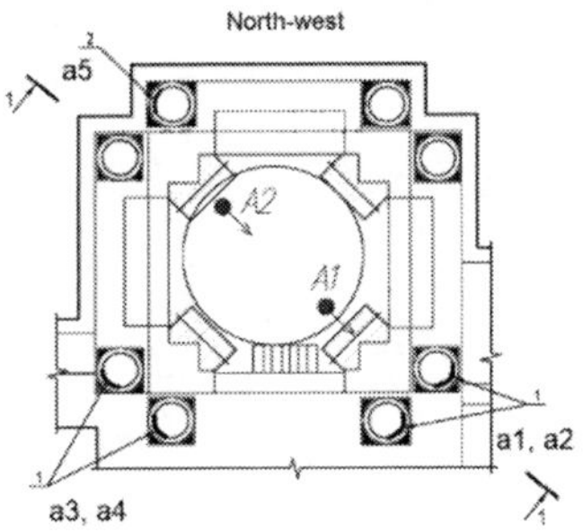

Fig. 4. Scheme of installing sensors for measuring vibrations and through probing (A1 and A2 – accelerometers. Digits 1 and 2 in figure designate places of brickwork probing – a1, a2, a3, a4, a5.

The measurement sensors have been arranged on the bearing structures. Results of frequency response analysis of vibrations are presented in Fig. 5 and 6.

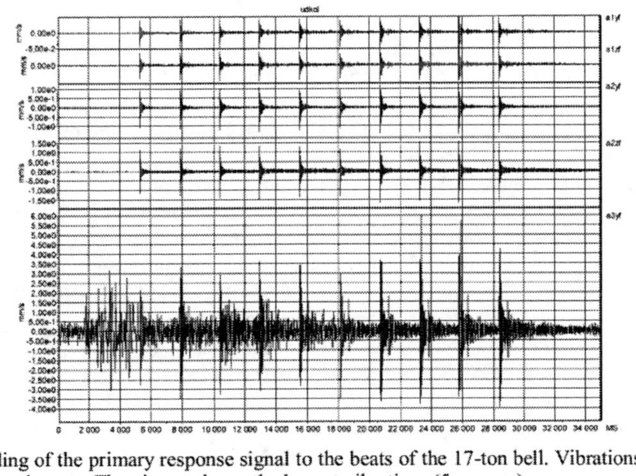

Fig. 5. Fragment of the recording of the primary response signal to the beats of the 17-ton bell. Vibrations of structures at the points of sensors installation are distinctly seen. The picture shows the largest vibrations (5 sensors).

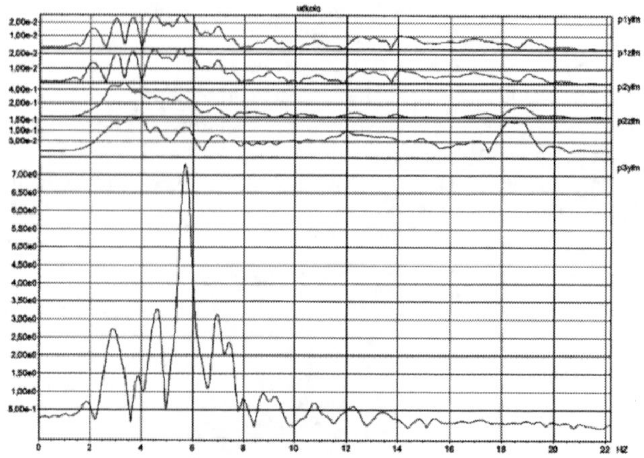

Fig. 6. Spectra obtained in the software signal processing environment with emphasizing of vibration peaks. Reaction to the strike of the 17-ton bell (short section). Sensor numbers are shown on the vertical axis to the right, on the horizontal axis – time in milliseconds. The amplitude of accelerations on each sensor, m / s^2 – on the vertical axis to the left.

The maximum registered velocity at sensors A3 and A4 (metal structures) amounted to 3 mm/s, and at sensors A1 and A2 (at brick pillars) – 1.5 mm/s. In this case displacements amounted to not more than 15 μm at the metal frame and 2.1 μm – in the brickwork (Fig. 7).

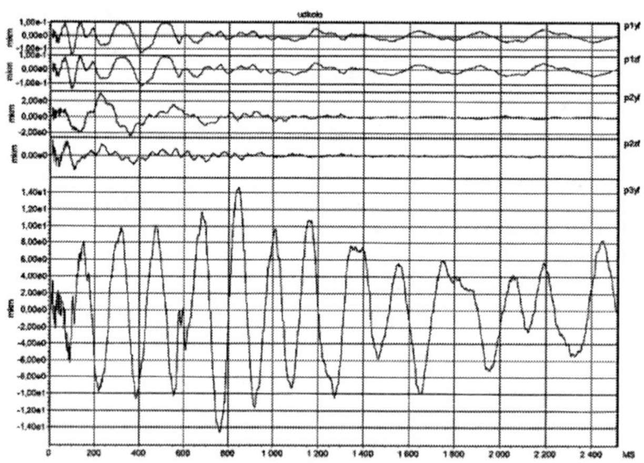

Fig. 7. Records of values when the structure is displaced along the horizontal axis, according to sensors, when a 17-ton bell is struck.

The analysis of amplitude of vibrations and spectra of oscillations of the system shows that the maximum values of acceleration and velocity in the range of frequencies posing a danger to the bearing structures (up to 50 Hz) do not exceed the values at the walls – up to 2.0 m/s^2, at that, the vibration velocity of the brick pillars does not exceed 0.9 mm/s, while the displacement is up to 2.5 μm.

The maximum registered velocity has amounted to 10 mm/s. In this case the displacements have not exceeded 15 μm at the metal frame and up to 1 μm – in the brickwork.

The values of accelerations from vibrations occurring during bell tower usage are lower than it is permitted by GOST sanitary standards (2010) for office premises for the steeple structures [5].

Apparently, it is difficult to expect any considerable vibrations at the structure base with the registered values of vibrations and displacements at the bell tower structures close to the attachment points of suspensions.

The settlements of the buildings foundations in weak soils reach 2 mm/year according to data of SN (1997) and TSN (2004) [6–7] under working conditions at the vibration velocities about 400 μm/s, while light damages of old buildings, separation of limewash, chipping of small pieces of plaster are possible at the vibration level of 1,200 μm/s.

The analysis of requirements of GOST (2007) [8] has made it possible to set a threshold value of vibration impact on the soil ground by the velocity of oscillations of 73 dB or 225 μm/s. In case this level is exceeded it will be necessary to carry out special engineering measures to reduce vibration negative consequences for maintaining the bearing capabilities of soils, reliable and trouble-free operation of engineering structures and provision of people's optimal vital activities. Even though, no registrations of soil oscillations at the base have been carried out while investigating vibration oscillations of the bell tower during chimes, it is apparent that the actual values of velocities and displacements are significantly below the threshold value of vibration impact, which can adversely affect the soil ground [9–10, 3].

Vibrations from ringing were recorded in the audio frequency range. Since the response of the structures to the external vibration effect from the bells and vibrations of the bells themselves was investigated, extremely low amplitudes of recorded vibrations were detected in the frequency range. Their detailed analysis revealed the inessential value of the Fourier spectra (Fig. 8).

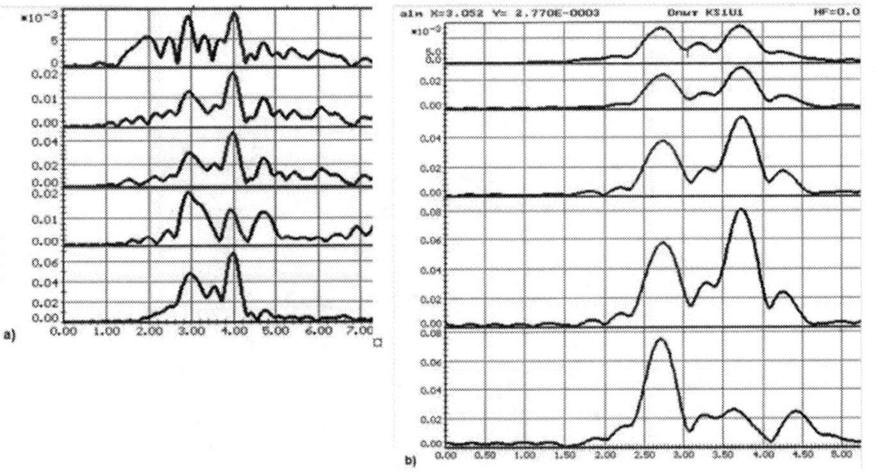

Fig. 8. Analysis of Fourier spectra in order to highlight resonant peaks corresponding to the level of vibrations and vibrations in a wide frequency range is given. Amplitude spectrum. The first tone (mode). Vertical range, transverse direction: a) 1 tone – 3.08 Hz (measurements in 2015); b) 1 tone – 2.7 Hz (measurements of 2016).

Since the largest peaks showed insignificant vibrations, the overlap of the Fourier spectra was not considered, due to the low values of vibrations (free vibrations) of the supporting structures and the foundation.

Thus the effect of the bell tower resonance swinging is not observed during multiple chimes, since the frequencies of clappers swinging and the natural frequency bending oscillations of the bell tower differ significantly.

The maximum oscillations in the course of regular work of bells occur in the steel frame structures. However, the pulse transmitted to the suspension beam and further to the bell tower is quite insignificant. The measurements of vibratory oscillations have shown that the maximum values of velocity and acceleration in the range of frequencies posing a danger to the bearing structures (up to 50 Hz) do not exceed the values at the walls – up to 2.0 m/s^2, at that, the vibration velocity of the brick pillars does not exceed 0.9 mm/s, while the displacements are up to 2.5 μm.

4. Discussion

Perhaps, further reasoning does not specifically emphasize the carried out analysis of vibration caused by the bells. However, it is necessary to take into account vibro-creeping (at least in the theoretical aspect) for estimating and calculating the sediment settlement under St. Isaac's Cathedral. This is relevant because there is a potential danger for the soil carrying capacity of the oscillations excited by the ringing of heavy bells.

The adverse impact sets in because of uneven settlement, when the soil becomes substantially non-uniform by its properties. It is important to take into consideration that the process of foundation settlement corresponds to quite a durable change, therefore, it is necessary to establish an instrumental monitoring and take the due engineering decisions for soils stabilization.

The vibration impact of chimes can weaken the bearing capability of soils at the bell tower base. It is common knowledge that the durable vibration loads on soil can bring about its nonreversible deformations, which will get gradually accumulated and can reach significant magnitudes. These deformations can bring about differential settlements of foundation, which will entail deformation of the elements of building structures, damage and destruction thereof.

Therefore, it is always desirable to look for the ways of reducing vibration loads on the bell towers. In this regard the heavy bells have been suspended in Christ the Saviour Cathedral using resilient suspensions with damping oscillations of the bells. In case of clapper verberation the bell skirt is in a position to move, which reduces the pulse being transferred from the strike to the brickwork of walls. They have chosen a different way in the Holy Trinity-St. Sergius Lavra. The specialists of Khrunichev Scientific and Production Center (division of Roskosmos (the Russian Federal Space Agency) have developed a clapper balanced with respect to the inertia moment for the Tsar Bell in the Kremlin of Moscow with such masses distribution that has many times reduced the pulse transmitted to the suspension beam and further, to the bell tower. The measurements of vibration oscillations have demonstrated that the amplitude of the bell tower horizontal oscillations has decreased 4 times with the balanced clapper as compared with a clapper of a traditional shape.

Generally, the bell towers, where the unique heavy bells are installed, are the famous historical sites. Historically, the shadoof suspension system was used originally in Russia, when the bell-clapper hung freely, while the bell swung together with the beam (shaft) of a suspension. The bell touched the clapper to produce reverberation chimes. A metal rod of a rectangular section, a tie beam, which ends have been forged to become round and put into bearings corresponding to wooden bars with a steel strip on the top established the basis of the shadoof bells suspension system. The suchlike system of bells ringing has been preserved in the Pskov Monastery of the Caves. The most ancient tie beam of early XVIth century (1508) has been preserved in the Kremlin; it is used for suspension of «Medved» (Bear) bell (i.e. «howler») of mezzanine of the Kremlin Ivan the Great steeple (Fig. 9 and 10):

Fig. 9. Shadoof bell in the «English style», i.e. bell swings full-circle. If in early times bells were swung through a small arc, but by the 17th century in England it was common to find bells capable of swinging through 360°. A short lever used earlier for swinging the bell, became essentially longer and took a shape of a big stick attached perpendicularly with one end to a rotating shaft with the bell. A rope had been tied to the other stick end, which was pulled by a chimer standing at the church wall. The foregoing stick came to be called the «shadoof».

Fig. 10. Shadoof «Medved» bell of the Kremlin Ivan the Great Bell Tower.

A passive position of the clapper with respect to the bell body in the shadoof suspension was quite relevant during installation of small bells. But with the increase of the shadoof suspension bell weight, they have been replaced with freely-hanging bells. Many bells, including small and medium ones were suspended in XV−XVII centuries statically to wooden beams embedded into pylons of bell towers and steeples alongside with the shadoof suspension of bells and the chime was produced by swinging the clapper.

Swinging the clapper, the chimers succeeded to achieve more durable sound. The suspension of bells is to be designed so that to provide high amplitude of sound pressure after clapper verberation. For example, the craftsmen in the Kremlin pulled the forged tie beams at suspensions through the pairs of side bell eyelets and secured the beams with the clamps in the form of U-letter to the suspension beam. A sturdy suspension provided for the freedom of the bell swinging from the verberation (chimes «by clappers»). After visual inspection of old suspensions it became clear that in the course of time the tie beams and clamps have ceased to provide a respective freedom of bell swinging. It was not difficult to come to a conclusion that significant bending moments affected the eyelets of bells during chimes (they can cause the destruction of eyelets and subsequent falling). In the course of bell swinging the vibration levels at frequencies normally about 2 Hz may cause significant destruction [11]. The multiple examples of bells, including the bells in the Kremlin steeple, prove particularly this mechanism of bells wreckage. Besides, the enslaving suspension becomes a reason for intensive transmission of oscillations of the chiming bell to the suspension beam and further, to the bell tower walls. A pulse from the clapper verberation and from vibration of the sounding bell is transmitted at every clapper verberation against the bell skirt through the bell suspension system and through the suspension beam to the bell tower. These pulses provoke resilient waves, which propagate along the brickwork of walls up and down and cause the bell tower oscillations.

It is known that the amplitude of the bell tower horizontal oscillations depends on the clapper mass. The bells foundry men have strictly followed a tradition: there should be a proportion of 1:25 between the mass of a clapper and a bell. This proportion should be proven by the practice of using bells, wherefore, not infrequently the cracks appeared in the skirt, when the clapper mass is bigger, if the clapper mass is smaller; it produces the basic tone of singing, i.e. the undertone of low sound level. It proceeds therefrom that it is important to plot a dependence of oscillations amplitude not on the mass of clapper, but on the mass of bell (under standard conditions the dependence of amplitude of bell tower displacements from the bell mass shall be linear). However, the amplitude of oscillations velocity and the amplitude of displacements in case of multiple verberations is approximately a couple of times bigger than in case of a single verberation. Therefore, when ringing the bell (clapper verberations against both sides), the amplitude of horizontal oscillations increases with the increase of quantity of verberations and attains a condition of «saturation» with 11-12 verberations. In our case, when investigating the north-western steeple of St. Isaac's Cathedral, the maximum velocities of oscillations, maximum displacements and the values of frequency of spectral maximum for horizontal oscillations in the north-western direction obtained at different steeple levels when ringing the bells have been measured. Using experimental data, it is also possible to build a dependence of amplitude of oscillations on elevation points. The acquired full-scale data can be described as a parabolic dependence proving the bending character of oscillations [12].

The analysis of bell tower oscillations under wind loads has shown that this frequency is a natural frequency of the steeple bending oscillations. The multiple verberations rather than a single one are relevant for the bell tower. Therefore, it is possible to clang all heavy bells simultaneously. At that, a superposition of oscillations excited by every bell takes place. According to investigations the amplitude of oscillations is higher in this case than any oscillation from any individual bell, but it is lower than the sum of amplitudes. It is explained by the fact that the periods of swings of the clappers of heavy bells are different with respect to values and the verberations proceed non-synchronously. GOST (2007) «Vibration and shock» establishes the limiting velocities of oscillations of the base of structures in case of impulse actions thereupon, which may include the verberations too. The GOST values significantly exceed the registered velocities of oscillations of the base of the north-western steeple of St. Isaac's Cathedral during chimes.

It is also important to involve a theory of cyclic strain-induced creep for assessing and calculating soil settlements. According to formula:

$$S = \alpha\sqrt{t} \ , \qquad (1)$$

where, S – settlements in mm, a – intensity of cyclic strain-induced creep, t – time in minutes. It equals $10^{-2} \div 10^{-1}$ mm/$\sqrt{\text{min}}$. for marshy soils with soil-bearing load of about 2 kg/cm^2 (such soil-bearing load is produced by the weight of the north-western steeple, including the cathedral base used for supporting it). This implies that the settlements can amount to 0.5 ÷ 0.05 mm during half an hour of chimes. In case of repeated impacts the settlements will get recommenced and summarized. A danger emerges not because of significant settlements, but rather due to irregularity owing to inhomogenuity of soil [10]. The obtained assessments of settlements show that they can not be ignored and it is reasonable to perform an annual leveling check of settlements of foundation or base of the bell tower.

5. Conclusions

In conclusion of the conducted investigations one can confirm that the vibrational impact of chimes of heavy bells with respect to the objects being considered can not adversely affect the bearing structures of the bell towers. This finding is proven out by already more than two years of successful operation of reconstructed objects. However, one should pay attention to the fact that the research of possible adverse consequences of bells installation works has been carried out already after changing «customary» conditions of operation. At that, there existed a probability of destruction or damage of the bearing structures as well as weakening of the base soils. Therefore, it is quite reasonable to recommend carrying out a definite type of investigations prior to beginning the restoration work: e.g., investigation of the bearing structures and building the mathematical model of the structure, which can be used for excluding possible emergency situations at the objects being reconstructed.

References

1. S. Savin & V. Tsakalidis. The use of elastic oscillations of different wavelengths to evaluate the dynamic parameters of buildings and structures and assess the strength of materials of the building construction. *COMPDYN 2015. Proceedings of the 5th International Conference on Computational Methods in Structural Dynamics and Earthquake Engineering Methods in Structural Dynamics and Earthquake Engineering, 5. 25-27 May 2015 Crete Island, Greece.* M. Papadrakakis, V. Papadopoulos, V. Plevris (eds.). Hersonissos, Crete: National Technical University of Athens. P. 706–720 (2015)

2. S.N. Savin. The use of elastic oscillations of different wavelengths to evaluate the dynamic parameters of buildings and structures and assess the strength of stone walls. Seysmostoykoye stroitel'stvo. Bezopasnost' sooruzheniy [Earthquake engineering. Constructions safety], **4**, 43–55 (2017)
3. S. Savin & E. Smirnova. Evaluation of mechanical safety of building structures using elastic vibrations varying in wavelength. World Applied Sciences Journal, **23**(11), 1448–1454 (2013) DOI: 10.5829/idosi.wasj.2013.23.11.13161
4. A.V. Adushkin, A.I. Goncharov, V.I. Kulikov. Vibrokolebanija kolokol'ni pri kolokol'nom zvone [The vibrations of bell tower excited by bell ringing]. Dinamicheskiye protsessy v geosferakh [Dynamic processes in geospheres], **1**, 172–180 (2010)
5. GOST (2010) Refractories and refractory raw materials. Methods of refractoriness determination – GOST R 53788-2010. Zdanija i sooruzhenija. Pravila obsledovanija i monitoringa tehnicheskogo sostojanija. Moscow: Standartinform (2010)
6. SN (1997). The sanitary norms of industrial vibration, vibration of residential and public buildings – SN 2.2.4/2.1.566-96. Proizvodstvennaya vibratsiya, vibratsiya v pomeshcheniyakh zhilykh i obshchestvennykh zdaniy. Minzdrav Rossii. Moscow: Minzdrav (1997)
7. TSN (2004). Designing the foundations of buildings and structures in St. Petersburg – TSN 50-302-2004. Proektirovanie fundamentov zdanij i sooruzhenij v Sankt-Peterburge. St. Petersburg, Strojizdat (2004)
8. GOST (2007) Vibration and shock. Vibration of buildings. Measurement of vibration and evaluation of its effects on structure – GOST R 52892-2007. Vibracija i udar. Vibracija zdanij. Izmerenie vibracii i ocenka ee vozdejstvija na konstrukciju. Moscow: Standartinform (2008)
9. A.D. Zhigalin & G.P Lokshin. Formirovanie vibracionnogo polja v geologicheskoj srede [The formation of the vibrational field in the geological sphere]. Inzhenernaja geologija [Engineering geology], **6**, 110–119 (1991)
10. A. Vel'sovskij, B. Karpov, E. Smirnova. Development of a new method for checking frost heave in roads. Proceedings of the ICE - Civil Engineering, **168**(5), 49–54 (2015). DOI: 10.1680/cien.14.00036
11. R. Smith, H. Hunt. Vibration of bell towers excited by bell ringing – a new approach to analysis. *Proceedings of the 23rd International Conference on Noise and Vibration Engineering, ISMA' 08, 9-15-2008 to 9-17-2008, Leuven, Belgium.* P. Sas (ed.). Heverlee (Belgium): Katholieke Univ. Leuven. P. 699–714 (2008)
12. A.J. Frost. *Towers and Bells.* 2nd edn. Guildford, Surrey: Central Council of Church Bell Ringers (2006)

On dynamic interaction between mechanical systems and selected electric motors

Tomasz Szolc[1,*], *Robert* Konowrocki[1], and *Andrzej* Pochanke[2]

[1]Institute of Fundamental Technological Research of the Polish Academy of Sciences, ul. Pawińskiego 5B, 02-106 Warsaw, Poland
[2]Warsaw University of Technology, Faculty of Electric Engineering, Plac Politechniki 1, 00-661 Warsaw, Poland

Abstract. In the paper there is presented a reliable structural model of the rotating mechanical systems as well as mathematical models of the stepping, synchronous and asynchronous motors, by means of which electromechanical coupling effects can be thoroughly investigated. An importance and severity of these phenomena, not sufficiently explored till present, have been demonstrated by results obtained for transient and steady-state operational conditions in the computational examples concerning torsional vibrations of drive trains with various electric motors.

1 Introduction

During last few decades electric motors became more and more commonly applied for driving of various machines, mechanisms and vehicles. Usually, such sources of power are characterized by simpler structures, they are more robust, efficient, precise to control and environmental friendly than e.g. piston and flow internal combustion engines or hydraulic motors. Nowadays, modern asynchronous, synchronous and stepping motors are particularly popular to drive mechanical objects with various geometrical dimensions, weights, rotational speeds and levels of power consumption. Because of this reason, electromechanical interactions with just these kinds of motors have been studied till present by many authors from numerous research centres in the whole world, e.g. in [1-5]. But still in majority of works mechanical engineers focus their attention on detailed dynamic properties of mechanical systems and they take into consideration an interaction of an electric motor in the form of 'a priori' assumed electromagnetic driving torque functions. Electrical engineers, however, very thoroughly model electric motors and treat the driven mechanical objects in very simplified forms of few degrees-of-freedom dynamic oscillators or even single rigid bodies. But by the use of such approaches many important electromechanical interaction effects occurring in steady-state and transient operational conditions cannot be considered. Namely, then it is very difficult to investigate an influence of motor rotor rotational speed fluctuations on oscillations of electric currents in the motor windings and, consequently, on fluctuations of the electromagnetic driving torque which is responsible for excitation of torsional vibrations of a mechanical system.

* Corresponding author: tszolc@ippt.pan.pl

In this paper dynamic interactions between mechanical systems in the form of rotating machines with working tools driven by electric motors by means of sets of stepped shafts, couplings and gear stages will be studied. As drive sources a steeping motor, permanent magnet synchronous motor (PMSM) and asynchronous motor are going to be considered.

2 Hybrid structural modelling of mechanical systems

In order to investigate a detailed character of the electromechanical coupling, the possibly realistic and reliable mechanical model of the driven system is applied. In this paper, similarly as e.g. in [6,7], dynamic investigations of the entire driven system are performed by means of the one-dimensional hybrid structural model consisting of finite continuous visco-elastic macro-elements and rigid bodies. In this model by the torsionally deformable cylindrical macro-elements of continuously distributed inertial-visco-elastic properties there are substituted successive cylindrical segments of the stepped shafts and coupling disks. In order to obtain a sufficiently accurate representation of the real object, the visco-elastic macro-elements in the hybrid model are characterized by the geometric cross-sectional polar moments of inertia responsible for their elastic and inertial properties as well as by the separate layers responsible for their inertial properties only. The inertias of gear-wheels and driven machine working tools are represented by rigid bodies attached to the appropriate macro-element cross-sections.

In the hybrid model torsional motion of cross-sections of each visco-elastic macro-element is governed by the local hyperbolic partial differential equations of the wave type. Mutual connections of the successive macro-elements creating the stepped shaft as well as their interactions with the rigid bodies are described by equations of boundary conditions. These equations contain geometrical conditions of conformity for rotational displacements of the extreme cross sections of the mutually adjacent elastic macro-elements as well as linear equations of equilibrium for external torques as well as for inertial, elastic and external damping moments.

The solution for forced vibration analysis has been obtained using the analytical – computational approach described e.g. in [6,7]. Solving the differential eigenvalue problem and an application of the Fourier solution in the form of series in the orthogonal eigenmode functions lead to the set of modal equations for time coordinates $\xi_m(t)$:

$$\ddot{\xi}_m(t)+\left(\beta+\tau\omega_m^2\right)\dot{\xi}_m(t)+\omega_m^2\xi_m(t)=\frac{1}{\gamma_m^2}\left(X_m^S\cdot T_{el}(t)-X_m^R\cdot M_r(t)\right),\quad m=1,2,\ldots \quad (1)$$

where ω_m are the successive natural frequencies of the drive system, β denotes the coefficient of external damping assumed here as proportional one to the modal masses γ_m^2, τ is the retardation time, $T_{el}(t)$ denotes the external torque generated by the electric motor, $M_r(t)$ is the driven machine retarding torque and X_m^S, X_m^R are the modal displacements scaled by proper maxima and corresponding respectively to the electric motor- and to the driven machine working tool-locations in the hybrid model. A fast convergence of the applied Fourier solution enables us to reduce the number of the modal equations to solve in order to obtain a sufficient accuracy of results in the given range of frequency. Here, it is necessary to solve only a few or at most a dozen or so modal equations (1), even in cases of very complex mechanical systems, contrary to the classical one-dimensional beam finite element formulation leading usually to large numbers of motion equations corresponding each to more than one hundred or many hundreds degrees of freedom (if the artificial and often error-prone model reduction algorithms are not applied).

3 Modelling of the electric motors

From the viewpoint of electromechanical coupling investigation properly advanced circuit models of the electric motors seem to be sufficiently accurate.

3.1 Modelling of the stepping motor

According e.g. to [5], the mathematical model of a typical four-cycle, double-phase stepping motor with the fundamental step angle 1.8 deg = 0.0314 rad, which means that its rotor is characterized by Z_r=50 salient poles, is described by two voltage equations:

$$L\frac{\mathrm{d}\,i_1(t)}{\mathrm{d}t}+Ri_1(t)-K_\mathrm{U}\frac{\mathrm{d}\,\Theta(t)}{\mathrm{d}t}\cdot\sin\left(\tilde{\Theta}(t)\right)=-U(t)\cdot\mathrm{sgn}\left\{\sin\left(\Phi(t)\right)\right\},\tag{2}$$

$$L\frac{\mathrm{d}\,i_2(t)}{\mathrm{d}t}+Ri_2(t)+K_\mathrm{U}\frac{\mathrm{d}\,\Theta(t)}{\mathrm{d}t}\cdot\cos\left(\tilde{\Theta}(t)\right)=U(t)\cdot\mathrm{sgn}\left\{\cos\left(\Phi(t)\right)\right\},\ \Phi(t)=\frac{\pi}{2}\int_0^t f(\tau)\mathrm{d}\tau,$$

where $i_1(t)$, $i_2(t)$ denote the electric currents in both motor phases, L is the phase inductance, R denotes the resistance of one phase, K_U is the motor voltage constant, $\Theta(t)$ denotes the instantaneous rotation angle of the rotor including the rigid body motion and the vibratory component, $U(t)$ is the slowly varying, controlled supply voltage, $\tilde{\Theta}(t)$ denotes the rotor electric angle, which in the case of the stepping motor can be determined as $\tilde{\Theta}(t)=Z_r\Theta(t)$, and $f(t)$ is the voltage supply commutation frequency. Here, a sufficiently good commutation realized by means of a proper stepping motor control should result in the control voltage supply phase angle $\Phi(t)\cong\tilde{\Theta}(t)$. The electromagnetic torque generated by the double-phase stepping motor is expressed by the following formula:

$$T_{el}(t)=K_\mathrm{T}\left[-i_1(t)\cdot\sin\left(\tilde{\Theta}(t)\right)+i_2(t)\cdot\cos\left(\tilde{\Theta}(t)\right)\right],\tag{3}$$

where K_T denotes the stepping motor torque constant. By substituting expression (3) into the modal equations (1) and upon a proper combinations of them with the voltage equations (2) one obtains the coupled set the parametric ordinary differential equations:

$$\mathbf{M}\ddot{\mathbf{r}}(t)+\mathbf{C}\left(\tilde{\Theta}(t)\right)\cdot\dot{\mathbf{r}}(t)+\mathbf{K}\left(\tilde{\Theta}(t)\right)\cdot\mathbf{r}(t)=\mathbf{F}\left(t,\dot{\mathbf{r}}(t)\right),\tag{4}$$

where: $\mathbf{C}\left(\tilde{\Theta}(t)\right)=\mathbf{C}_0+\mathbf{C}_\mathrm{E}\left(\tilde{\Theta}(t)\right)$, $\quad\mathbf{K}\left(\tilde{\Theta}(t)\right)=\mathbf{K}_0+\mathbf{K}_\mathrm{E}\left(\tilde{\Theta}(t)\right)$,

$$\mathbf{r}(t)=\mathrm{col}\left[i_1(t),i_2(t),\xi_0(t),\xi_1(t),\xi_2(t),\ldots\right].$$

The symbols **M**, $\mathbf{C}_0$ and $\mathbf{K}_0$ denote, respectively, the constant diagonal modal mass, damping and stiffness matrices, $\mathbf{C}_\mathrm{E}(\tilde{\Theta}(t))$ is the band matrix of the inductive-electro-magnetic effects and $\mathbf{K}_\mathrm{E}(\tilde{\Theta}(t))$ denotes the band matrix of the resistant-electro-magnetic effects, both of harmonically variable coefficients with the frequency following from the current electric rotation angle. The symbol $\mathbf{F}(t,\dot{\mathbf{r}}(t))$ denotes the external excitation vector due to the control input voltage and the retarding torques. The unknown co-ordinate vector $\mathbf{r}(t)$ consists of the electric currents in both motor phases and of the unknown time functions $\xi_m(t)$ in the Fourier solutions, m=0,1,2,.... In order to obtain the system's dynamic response, equations (4) are solved by means of a direct integration. The number of equations (4) corresponds to the number of eigenmodes taken into consideration in the range of frequency of interest. These equations are mutually coupled by the parametric terms expressing the electromagnetic interaction with the stepping motor and they are very convenient for effective numerical simulations of a dynamic behaviour of the electromechanical system.

3.2 Modelling of the permanent magnet synchronous motor (PMSM)

In the case of a permanent magnet symmetrical, 3-phase synchronous motor with a uniform air-gap distribution and stator windings star-connected, voltage equations formulated in the natural system of electrical exes "a-b-c" have the following form:

$$L\frac{\mathrm{d}\,i_a(t)}{\mathrm{d}t}+M\frac{\mathrm{d}\,i_b(t)}{\mathrm{d}t}M\frac{\mathrm{d}\,i_c(t)}{\mathrm{d}t}+Ri_a(t)-\psi\frac{\mathrm{d}\tilde{\Theta}(t)}{\mathrm{d}t}\cdot\sin\left(\tilde{\Theta}(t)\right)=U(t)\sin\left(\tilde{\Theta}(t)\right), \tag{5}$$

$$L\frac{\mathrm{d}\,i_b(t)}{\mathrm{d}t}+M\frac{\mathrm{d}\,i_a(t)}{\mathrm{d}t}+M\frac{\mathrm{d}\,i_c(t)}{\mathrm{d}t}+Ri_b(t)-\psi\frac{\mathrm{d}\,\tilde{\Theta}(t)}{\mathrm{d}t}\sin\left(\tilde{\Theta}(t)-\frac{2}{3}\pi\right)=U(t)\sin\left(\tilde{\Theta}(t)-\frac{2}{3}\pi\right),$$

$$L\frac{\mathrm{d}\,i_c(t)}{\mathrm{d}t}+M\frac{\mathrm{d}\,i_a(t)}{\mathrm{d}t}+M\frac{\mathrm{d}\,i_b(t)}{\mathrm{d}t}+Ri_c(t)-\psi\frac{\mathrm{d}\,\tilde{\Theta}(t)}{\mathrm{d}t}\sin\left(\tilde{\Theta}(t)+\frac{2}{3}\pi\right)=U(t)\sin\left(\tilde{\Theta}(t)+\frac{2}{3}\pi\right),$$

where $i_a(t)$, $i_b(t)$), $i_c(t)$ denote the electric currents in motor phases, M denotes the mutual stator coil inductance and ψ is the flux generated by the permanent magnet. The remaining symbols have been already defined above. Upon the similar combination as that performed in the case of the stepping motor one obtains the analogous to (4) coupled set the parametric ordinary differential equations. In order to carry out qualitative analyses, using Clark-Park's transformation for the currents and voltages in Eqs. (5), one obtains:

$$\begin{bmatrix} R+(L-M)\frac{\mathrm{d}}{\mathrm{d}t} & (L-M)\frac{\mathrm{d}\tilde{\Theta}(t)}{\mathrm{d}t} & 0 \\ -(L-M)\frac{\mathrm{d}\tilde{\Theta}(t)}{\mathrm{d}t} & R+(L-M)\frac{\mathrm{d}}{\mathrm{d}t} & -\psi\frac{\mathrm{d}\tilde{\Theta}(t)}{\mathrm{d}t} \end{bmatrix}\cdot\begin{bmatrix} i_d \\ i_q \\ 1 \end{bmatrix}=\begin{bmatrix} u_d \\ u_q \end{bmatrix}. \tag{6}$$

The above equations make the coupling between the electrical and mechanical part nonlinear. Upon Clark-Park's transformation the same property has a circuit model of the asynchronous motor. Thus, according to [4], qualitative analyses carried out for both kinds of electric motors can lead to a determination of rotor-to-stator electromagnetic rotational stiffness- and damping characteristics using the harmonic balance method to solve analytically the voltage equations transformed into mutually perpendicular electric axes.

3.3 Modelling of the asynchronous motor

In the case of the symmetrical three-phase asynchronous motor electric current oscillations in its windings are described by the six circuit voltage equations transformed next into the system of four Park's equations in the so called 'αβ-*dq*' reference system, [4]:

$$\begin{bmatrix} \sqrt{\frac{3}{2}}U\cos(\omega_e t) \\ \sqrt{\frac{3}{2}}U\sin(\omega_e t) \\ 0 \\ 0 \end{bmatrix}=\begin{bmatrix} L_1+\frac{1}{2}M & 0 & \frac{3}{2}M & 0 \\ 0 & L_1+\frac{1}{2}M & 0 & \frac{3}{2}M \\ \frac{3}{2}M & 0 & L_2'+\frac{1}{2}M & 0 \\ 0 & \frac{3}{2}M & 0 & L_2'+\frac{1}{2}M \end{bmatrix}\cdot\begin{bmatrix} \dot{i}_\alpha^s(t) \\ \dot{i}_\beta^s(t) \\ \dot{i}_d^r(t) \\ \dot{i}_q^r(t) \end{bmatrix}+ \tag{7}$$

$$+\begin{bmatrix} R_1 & 0 & 0 & 0 \\ 0 & R_1 & 0 & 0 \\ 0 & \frac{3}{2}pM\frac{\mathrm{d}\Theta(t)}{\mathrm{d}t} & R_2' & p\frac{\mathrm{d}\Theta(t)}{\mathrm{d}t}\left(L_2'+\frac{1}{2}M\right) \\ -\frac{3}{2}pM\frac{\mathrm{d}\Theta(t)}{\mathrm{d}t} & 0 & -p\frac{\mathrm{d}\Theta(t)}{\mathrm{d}t}\left(L_2'+\frac{1}{2}M\right) & R_2' \end{bmatrix}\cdot\begin{bmatrix} i_\alpha^s(t) \\ i_\beta^s(t) \\ i_d^r(t) \\ i_q^r(t) \end{bmatrix},$$

where ω_e is the supply voltage circular frequency, L_1, L_2' are the stator coil inductance and the equivalent rotor coil inductance, respectively, M denotes the relative rotor-to-stator coil inductance, R_1, R_2' are the stator coil resistance and the equivalent rotor coil resistance, respectively, p is the number of pairs of the motor magnetic poles and i_α^s, i_β^s are the electric currents in the stator windings reduced to the electric field equivalent axes α and β and i_d^r, i_q^r are the electric currents in the rotor windings reduced to the electric field equivalent axes d and q, [4]. Then, the total electromagnetic torque and its oscillatory part generated by such a motor can be expressed by the following formulae:

$$T_{el} = \frac{3}{2} pM\left(i_\beta^S \cdot i_d^r - i_\alpha^S \cdot i_q^r\right), \qquad T_{el}^{\mathrm{var}}(t) = S(\omega)\cdot\sin(\omega t) + T(\omega)\cdot\cos(\omega t). \tag{8}$$

By projecting the sine- $S(\omega)$ and cosine- $T(\omega)$ components of the electromagnetic torque oscillatory part as well as the analogous components of the rotor rotation angle, respectively, on the complex plane real and imaginary axes and using the proper definitions given e.g. in [3], the electromagnetic torsional stiffness $k_e(\omega)$ and the coefficient of damping $d_e(\omega)$ generated by the asynchronous motor are determined in the following form:

$$k_e(\omega) = -\frac{U\cdot S(\omega) + W\cdot T(\omega)}{U^2 + W^2}, \qquad d_e(\omega) = -\frac{1}{\omega}\cdot\frac{U\cdot T(\omega) - W\cdot S(\omega)}{U^2 + W^2}, \tag{9}$$

where: $$W = \sum_{m=0}^{\infty} \frac{(X_m^S)^2 T(\omega)\cdot(\omega_m^2 - \omega^2) - \left[(X_m^S)^2 S(\omega) - X_m^S X_m^R R\right]\cdot(\beta + \tau\omega_m^2)\omega}{\gamma_m^2\left[(\omega_m^2 - \omega^2)^2 + (\beta + \tau\omega_m^2)^2\omega^2\right]},$$

$$U = \sum_{m=0}^{\infty} \frac{\left[(X_m^S)^2 S(\omega) - X_m^S X_m^R R\right]\cdot(\omega_m^2 - \omega^2) + (X_m^S)^2 T(\omega)\cdot(\beta + \tau\omega_m^2)\omega}{\gamma_m^2\left[(\omega_m^2 - \omega^2)^2 + (\beta + \tau\omega_m^2)^2\omega^2\right]}.$$

The above expressions derived in the framework of a qualitative analysis of properties of the asynchronous motor, that drives a linear model of the rotating machine, enable us to carry out thorough dynamic investigations of the coupled electromechanical system.

4 Computational examples

In the first computational example there is performed a simulation of the run-up, steady state operation and run-down of the precise geared drive system driven by the stepping motor. A mechanical model of this object is shown in Fig. 1. It has been uniformly accelerated from its standstill to the constant average rotational speed n=210 rpm and nominal torque 0.35 Nm within 3 s in order to operate for next 1 s under the constant retarding torque generated by the power receiver. Then, within successive 3 s the drive system was uniformly stopped back to the standstill. In Fig. 2 in time and frequency domain there are presented plots of the considered system dynamic response obtained for two simulation modes: for the "coupled mode", where the results have been determined by means of an integration of Eqs. (4) and marked by the black lines, as well as for the traditionally applied "uncoupled mode" illustrated here by the grey lines. In the latter case the mechanical system was treated as a one rigid body, i.e. only modal co-ordinate ξ_0 in (4) has been taken into consideration in order to determine the motor electromagnetic torque (5) to be imposed next as an external excitation of the hybrid mechanical model of the drive system. It is worth noting that the results obtained for these two modes are significantly different from each other both in transient and steady-state operational conditions. During the start-up and run-down the rotational speeds as well as the dynamic torques observed in

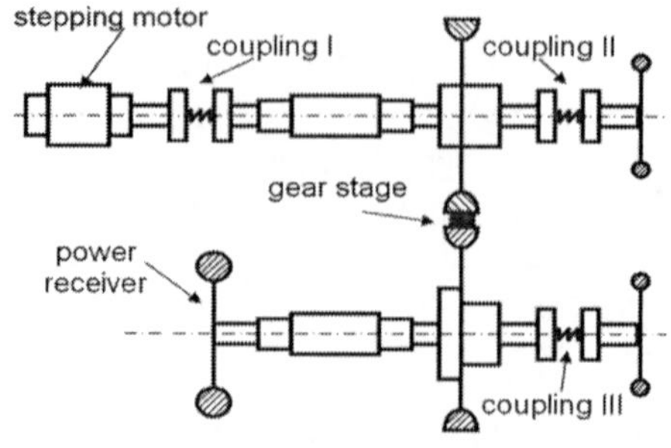

Fig. 1. The hybrid mechanical model of the precise drive system.

the system input and output shaft are characterized by severe transient resonances with the fundamental eigenmode of the first natural frequency equal 103.4 Hz, as shown in Figs. 2b-d. The amplitudes of this resonances are particularly severe in the case of the uncoupled mode, although in these operational conditions the coupled mode yields much more severe external excitation produced by the driving motor, which follows from Fig. 2a. However, during the steady-state operation in the case of uncoupled mode

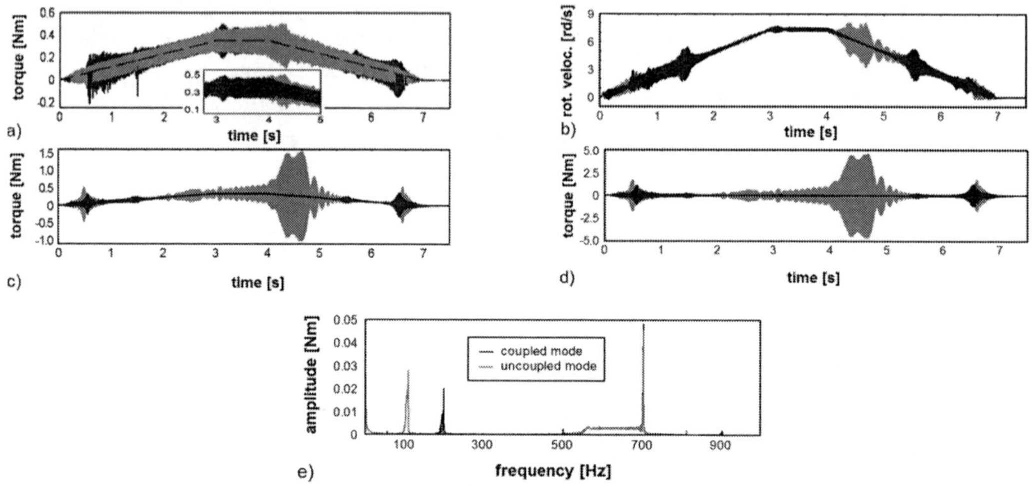

Fig. 2. Dynamic response of the electromechanical system: time histories of the electromagnetic motor torque (a), rotational speed of the output shaft (b), dynamic torques transmitted by the input (c) and output (d) shaft, and the amplitude spectra of the stepping motor torque (e).

the abovementioned regular and slightly stronger electromagnetic excitation results in resonant responses both for the rotational speeds and the dynamic torques. In order to explain this fact the FFT analysis of the time-histories of the excitation torque generated by the stepping motor has been performed for the coupled and uncoupled mode. The amplitude spectra obtained for the coupled and uncoupled mode and depicted in Fig. 2e are characterized by almost identical the greatest peaks of frequency 700 Hz corresponding to the fundamental excitation component produced by the stepping motor in the steady-state operation conditions. This value results from the assumed nominal rotational speed n=210 rpm and the fundamental step angle 0.0314 rad of the considered motor. In addition to these peaks with frequency of 700 Hz, different sub-harmonic components of the excitation torque are generated. In the case of uncoupled mode, in Fig. 2e the significant excitation peak of frequency ca. 100 Hz is observed, which is very close to the first drive system natural frequency equal to 103.4 Hz. This component is responsible for the observed resonance effects depicted in Figs. 2b-d for the uncoupled mode, in a contradistinction to the coupled mode, where the analogous sub-harmonic excitation component is characterized by frequency ~200 Hz which is far away from the first and the second system natural frequency equal to 344.4 Hz. According to the above, one can conclude that the taken into consideration vibratory, inertial-visco-elastic properties of the mechanical system essentially influence qualitatively and quantitatively the electromechanical coupling effects resulting respectively in different dynamic responses.

In the second computational example the dynamic investigations will be carried out for a rotor machine with a typical structure for blowers, compressors, pumps or high-speed beater mills. Namely, their rotor-shaft contains a heavy impeller attached to the drive shaft in an overhung form and is driven by the 200 kW, 400 V Y asynchronous motor with the nominal rotational speed of 2980 rpm through a flexible coupling with a non-linear characteristic. There are also worth noting the characteristic features of such device, i.e. a huge polar mass moment of inertia of the whole driven part in comparison with the polar mass moment of inertia of the motor rotor as well as a much smaller torsional stiffness of the elastic coupling than torsional stiffness values of all rotor-shaft segments. The hybrid model of this object is presented in Fig. 3.

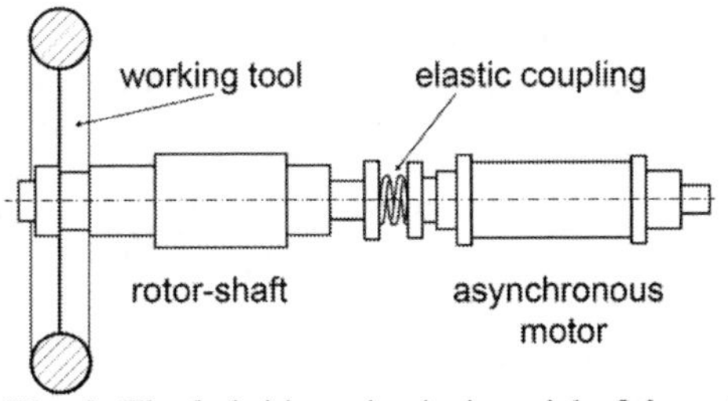

Fig. 3. The hybrid mechanical model of the rotor machine drive system.

In Fig. 4a there are shown static characteristics of the considered asynchronous motor determined according to [4] for four variants of the starting torque values which depend on the rotor resistance: namely, the greater the resistance, the greater the starting torque. In Fig. 4b characteristics of the electromagnetic stiffness and coefficients of damping generated by the considered asynchronous motor within the torsional interaction frequency range of 0-100 Hz are plotted. In this figure the thin lines correspond to the stiffness characteristics and the thick lines, appropriately, to the damping ones. These characteristics have been determined by means of relationships (9). Using respectively the same colours, these plots correspond to the mentioned above four variants of rotor resistances and thus to four variants of motor static characteristics presented in Fig. 4a. It is to emphasize that all the electromagnetic damping characteristics indicate negative value zones. From Figs. 4a and 4b it follows that for the considered here four starting motor torque variants the breadths of these negative damping zones gradually decrease with the rise of the starting motor torque values.

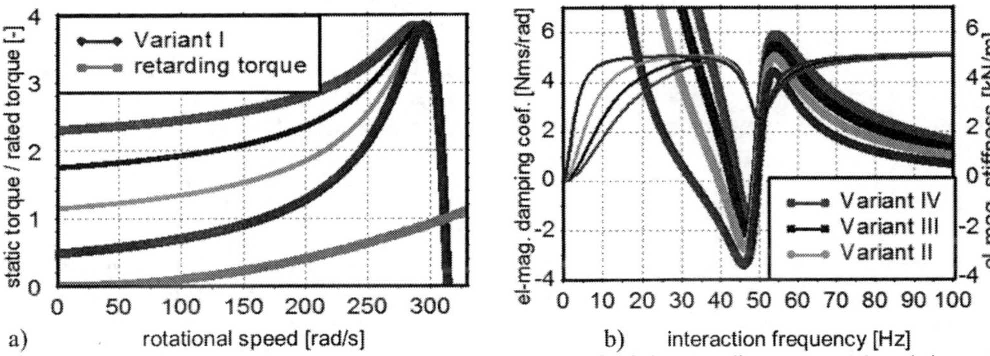

Fig. 4. Static characteristics of the asynchronous motor and of the retarding torque (a) and dynamic characteristics of the rotor-to-stator electromagnetic damping and stiffness (b).

In order to demonstrate a danger created by the negative electromagnetic damping generated by the asynchronous motor, start-ups of the considered rotor machine to the rotational speed of 2400 rpm within 60 s have been performed. Then, for the aim of examination of a sensitivity of the considered drive train to operational instabilities, after next 12 s of the rotor machine steady state motion a slight step-wise increase of the retarding torque was assumed. For such an operation scenario the run-up simulation results are depicted in Fig. 5a-b. In the face of commonly observed very weak structural damping

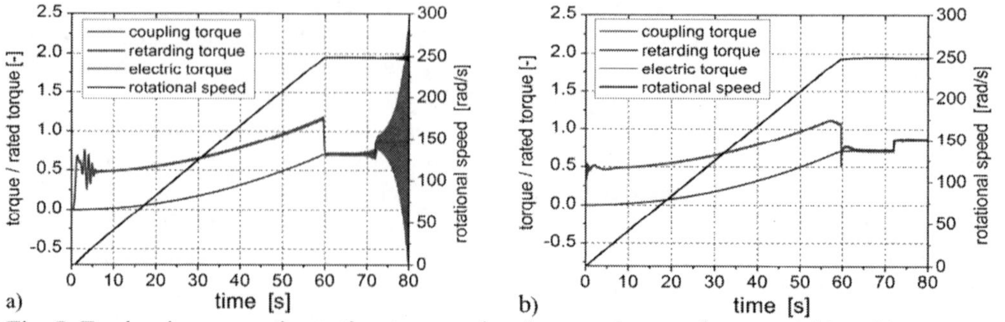

Fig. 5. Torsional response due to the start-up using the open-loop scalar control (a) and by means of the closed-loop PID vector control (b) of the starting motor torque.

in torsionally vibrating systems an application of the open-loop scalar control '*U*/*f*=const' of the asynchronous motor results in instability caused by the negative electromagnetic damping, as shown in Fig. 5a. Here, the rotational speed 2400 rpm corresponds to a synchronous excitation of 40 Hz which is contained in the negative damping zone of the motor starting torque Variant I depicted in Fig. 4b. Then, due to the existing deficiency of resultant positive damping in the system, free torsional vibrations with its first natural frequency of 34.7 Hz and rapidly increasing amplitudes are induced. If in the considered system the negative electromagnetic damping cannot be naturally compensated by the mechanical one, a proper closed-loop control of the asynchronous motor is able to prevent such instability. In Fig. 5b there are presented time-histories of a dynamic response obtained for exactly the same system parameters as before, but for the PID control realized by means of vector inverter with gains determined using Ziegler-Nichols' or Cohen-Coon's methods. The both approaches result in the visible stabilization of the considered object.

5 Final remarks

The performed study confirms that dynamic interaction between electric motors and mechanical systems can be thoroughly investigated when sufficiently reliable physical and mathematical models of these objects are applied. From the computational results it follows that torsional vibrations of the driven mechanical system essentially influence qualitatively and quantitatively its excitation by the stepping motor. Furthermore, negative damping generated by the asynchronous motor is responsible for operational instability of the entire drive train and by means of a proper motor control this dangerous effect can be eliminated.

References

1. B.F. Evans, A.J. Smalley, H.R. Simmons, ASME Paper, **85**-DET-122, (1985)
2. A. Tabesh, R. Iravani, IEEE Trans. Energy Convers. **20** (2), 268–275, (2005)
3. T.P. Holopainen, A.-K. Repo, J. Järvinen, *Proc. of the 8th IFToMM Int. Conf. on Rotordynamics* (KIST, Seoul, Korea, 2010), 986–993.
4. T. Szolc, R. Konowrocki, M. Michajłow, A. Pręgowska, MSSP, **49**, 118-134, (2014)
5. R. Konowrocki, T. Szolc, A. Pochanke, A. Pręgowska , MSSP, **70-71**, 397-413, (2016)
6. T. Szolc, *Int. J. Rot. Mach.* **6** (2), 135-149, (2000)
7. T. Szolc, *A dynamic analysis of complex discrete-continuous mechanical systems.* Habilitational dissertation (IFTR Reports, Warsaw 2/2003) *(in Polish)*

MATEC Web of Conferences **285**, 00019 (2019) https://doi.org/10.1051/matecconf/201928500019
DYNKON 2019

Vibration damping in road bridges – two case studies of full scale experiments

Daniel Szynal[1,*], and *Lucjan* Janas[1]

[1] Rzeszow University of Technology, The Faculty of Civil and Environmental Engineering and Architecture, 35-959 Rzeszów, Poland

Abstract. The paper presents the results of a vibration damping analysis of two road viaducts. The analysis was based on displacement signals recorded during the dynamic load tests carried out before an acceptance process of the viaducts. To create an additional excitation, a 5 cm high threshold was located on the road during low-velocity passages. The Logarithmic Decrement (LD) was calculated using an exponential curve fit to free vibration amplitudes. A fundamental frequency of vibration was considered. To minimize the influence of signal imperfections, a filtration process was carried out using bandpass Infinite Impulse Response (IIR) Bessel filter. The results were correlated for various degrees of excitation resulting from different velocities of loading vehicles and the presence of the threshold. Examples of appropriate and inadequate signals for the damping analysis were shown.

1 Introduction

Damping is one of the major issues describing dynamic behaviour of a structure. In contrast to other parameters, like mass and stiffness, the knowledge about it is still limited. It is obvious that a bigger damping level is better for a structure. Damping has an impact on safety, durability, and exploitation comfort of a structure. Too low damping can cause excessive vibrations that can affect the durability of the structure and the comfort of use – especially in light structures. In some cases a special damping devices have to be installed to decrease vibrations of the structure.

Bridge standards and regulations do not define the particular requirements for damping level. Lower limits of damping ratio for calculation purposes in the railway bridges dynamic analysis are presented in the standard [1]. It is also noted that effects of maximum dynamic load (peak response of a traffic speed corresponding to resonant loading) are highly dependent upon damping.

The precise evaluation of damping during a design phase is practically impossible. Estimating formulas for damping level in footbridges are known [2], although calculated

* Corresponding author: d.szynal@prz.edu.pl

values are marked with high uncertainty. For this reason, many experimental researches of damping level in footbridges are being made [3, 4, 5, 6]. The case is more complicated for road bridges. Due to a development of these structures, the issue acquires significance.

The damping phenomenon depends on many factors including type of a structure, degree of freedom, static scheme, types of materials, connections, and bearings. Furthermore, damping cannot be determined as a parameter, which permanently and univocally typify a relevant structure. A damping level varies depending on a vibration frequency, temperature or degree of vibration excitation. On this account, damping cannot be described by the simple physical equations used in the civil engineering. During a design phase it is advised to rely on the existing structure experiences gained through experimental tests.

This paper presents a damping research of two road viaducts. The analysis was based on displacement signals recorded during dynamic load tests.

2 Measures expressing a level of damping

The most popular measure expressing a level of damping, mainly because of its simplicity, is the *Logarithmic Decrement* (LD). There are several methods of computing provided it in the literature. One of the basic and most commonly used is the equation (1):

$$\delta = \frac{1}{m} ln \frac{A_n}{A_{n+m}} \tag{1}$$

where A_n and A_{n+m} signify value of the first and last amplitude of analysed signal, respectively.

Usually, ten first amplitudes are considered [7]. In this paper, LD was calculated by a method basing on the exponential curve fit to vibration amplitudes, which is recommended by the standard [8] and widely described in [7], among others. After determining the parameters of the curve depicted by the equation (2), LD (δ) is computed as a multiplication of β and period T as shown in equation (3):

$$y = A_0 e^{-\beta x} \tag{2}$$

$$\delta = \beta \cdot T \tag{3}$$

where A_0 signifies value of the initial amplitude.

Another commonly used measure of damping is the *damping ratio*. A relationship between LD and damping ratio is expressed by the equation (4):

$$\zeta = \frac{\delta}{2\pi} \tag{4}$$

3 Test procedure and signal analysis

The structure displacement signals during dynamic load test were measured by inductive transducers with the 0,01 mm resolution. The measuring equipment was calibrated in an accredited laboratory. The test procedure was verified by a comparative research [9]. The vibration signals were recorded with the 200 Hz frequency. Several passages of the loading vehicle at 10, 30, 40, and 50 km/h on an even road (fig. 1a), and a passage over the 5 cm high threshold at 5 or 10 km/h were recorded to implement an additional excitation [10] (fig. 1b). Because of the shape of the road near the viaduct, higher velocities could not be reached.

a) b)

Fig. 1. A passage of the loaded truck during a dynamic load test: a) on the even road, b) over the 5 cm high threshold.

The signals suitable for the damping analysis were separated from the results of field test. Usefulness of the signal depends on several factors, including a degree of vibration excitation, level of noise or number of modal shapes excited during a passage. Since the calculated values of LD cannot be compared and verified with any other sources, it is necessary to choose the signals with least probability of inaccuracy. In the damping analysis, only free vibration of the structure is considered, i.e. when the last axle of the vehicle leaves the structure. Then, no additional sources of excitation resulting from the truck influence are present. Furthermore, a presence of the vehicle mass on the structure causes an excess damping factor, which would inflate the value of LD. After the extraction of free vibration signal, its suitability for the damping analysis is assessed. The beginning of free vibration was determined based on the time of passage on the viaduct.

Recorded signals are rarely suitable for the damping analysis without prior processing because of the occurring noise and interference or excitation of more than one modal shape. These issues affect the accuracy of exponential curve fit. To minimize the influence of signal imperfections, a filtration process was carried out. In this paper, a bandpass type infinite impulse response (IIR) filter was applied. After detrmining the fundamental frequency by analysing the vibration spectrum, a lower and upper cut-off frequencies were set. A vibration beyond the defined scope is separated from the signal. During the process of filtration, the signal will unavoidably change due to phenomena like a phase lag or lowering the values of initial amplitudes. The level of deformation depends on the type of applied filter, its order, and bandwidth. The higher row and narrower bandwidth, the greater deformations in the output signal. In the case of LD analysis, it is recommended to use the Bessel filter [11]. The analysis of two road viaducts presented in this paper was carried out using filter bandwidth broader than 2 Hz and order not higher than 5.

4 Results of research

4.1 W-1 beam viaduct

The first of analysed structures, marked as W-1, is a two-span beam viaduct with a continuous beam static scheme (fig. 2). Both spans are 30 m long. The superstructure is designed as two post-tensioned trapezium girders monolithically connected with a reinforced concrete deck slab. The substructure consists of U-shaped abutments and a pier made of two separate columns, one under each girder. The viaduct is at an angle of 84.11° to the highway below it. The span structure is supported on the substructure by means of elastomeric bearings. The supports are bedded on shallow foundations made of reinforced concrete footings. The viaduct was designed in compliance with the A class according to the standard [12]. The W-1 viaduct is a typical structure in case of modern bridge structures currently built above highways.

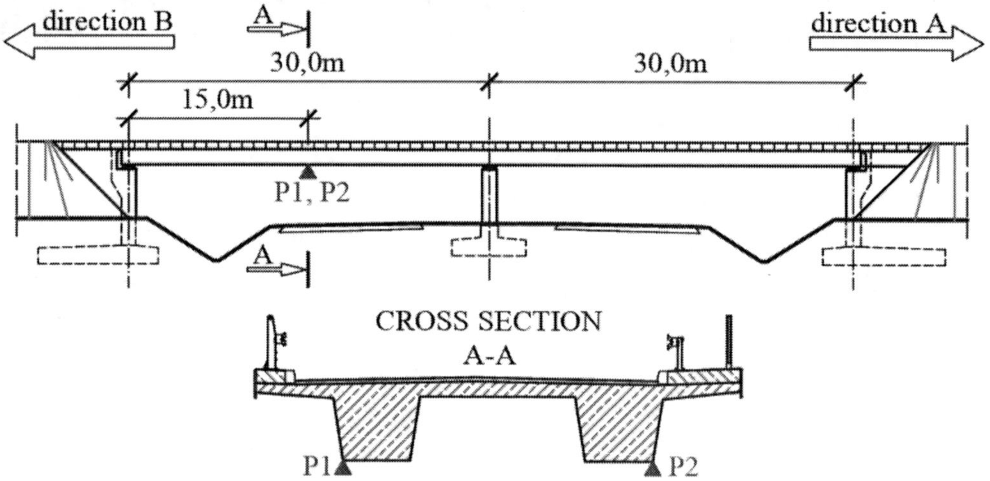

Fig. 2. The construction scheme of the W-1 viaduct with indicated measurement points.

Six signals selected for the analysis had at least ten amplitudes with the values no less than resolution of the inductive transducers used during the field test, i.e. 0.01 mm. The figures 3a and 3b present the signal inadequate for the damping analysis because of the low excitation, and the appropriate signal with relatively high excitation, respectively.

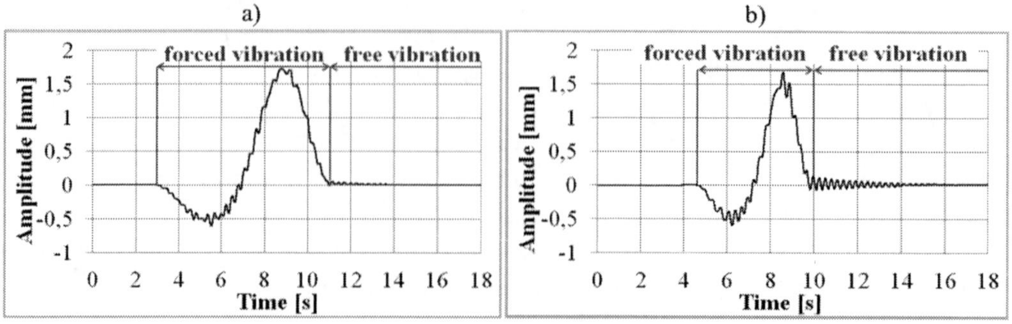

Fig. 3. The examples of measured signals: a) low excitation, inadequate for the damping analysis b) relatively high excitation – appropriate for the damping analysis.

After extracting the free vibration range, the signal was filtrated. From the processed signal, the values of successive amplitudes were determined. The exponential curve was fit by means of a least-squares method excluding the initial amplitudes of values lowered by the filtration process. After determining the exponential curve parameters and period of vibration, LD was calculated. The figures 4a and 4b present the filtered vibration with curve fit, and vibration spectrum, respectively.

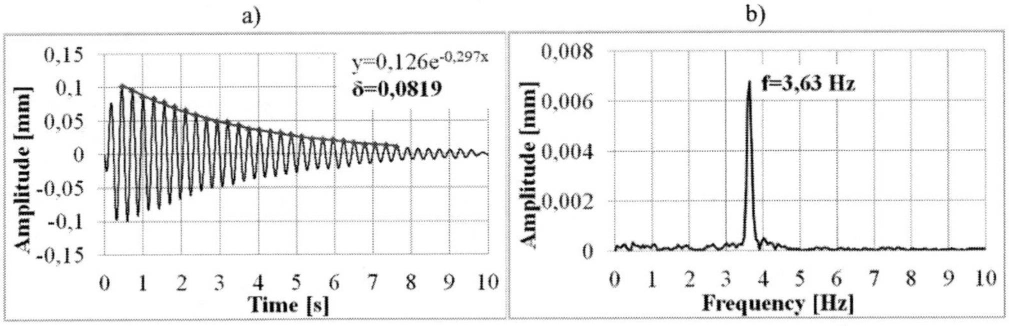

Fig. 4. The vibration damping analysis for a passage at a velocity of 40 km/h in the B direction: a) free vibrations and exponential curve fit to the vibration amplitudes, b) vibration spectrum.

The calculated values of LD and damping ratio for vibration with the fundamental frequency f=3,63 Hz are presented in table 1. The mean value of LD is 0.088 and matches the *average damping* according to [11], which is typical for most bridges and viaducts $(0.05 < \delta < 0.15)$.

Table 1. The values of LDT and damping ratio for W-2 viaduct.

Velocity/direction of passage	Measurement point	Initial amplitude [mm]	LD	Damping ratio
30/A	P1	0,015	u.c.	u.c.
	P2	0,013	u.c.	u.c.
40/A	P1	0,021	u.c.	u.c.
	P2	0,020	u.c.	u.c.
40/B	P1	0,019	u.c.	u.c.
	P2	0,022	u.c.	u.c.
50/B	P1	0,101	0,097	0,015
	P2	0,087	0,082	0,013
10/A (threshold)	P1	0,041	0,082	0,013
	P2	0,043	0,087	0,014
10/B (threshold)	P1	0,101	0,099	0,016
	P2	0,970	0,083	0,013
		Mean value:	**0,088**	**0,014**
		Standard deviation:	0,007	0,001

u.c. – unable to calculate LD due to the low excitation of the construction

4.2 W-2 frame viaduct

The second analysed viaduct, marked as W-2, is a three-span frame structure (fig. 5). The structure of side spans is designed as a post-tensioned concrete slab. The middle span is made of two post-tensioned trapezium girders monolithically connected with a reinforced concrete deck slab. The viaduct is shaped at an angle of 62.1° towards the obstacle. The soil under the shallow foundation of the structure was strengthened by means of the DSM technology. The viaduct was designed in compliance with the A class according to the standard [12].

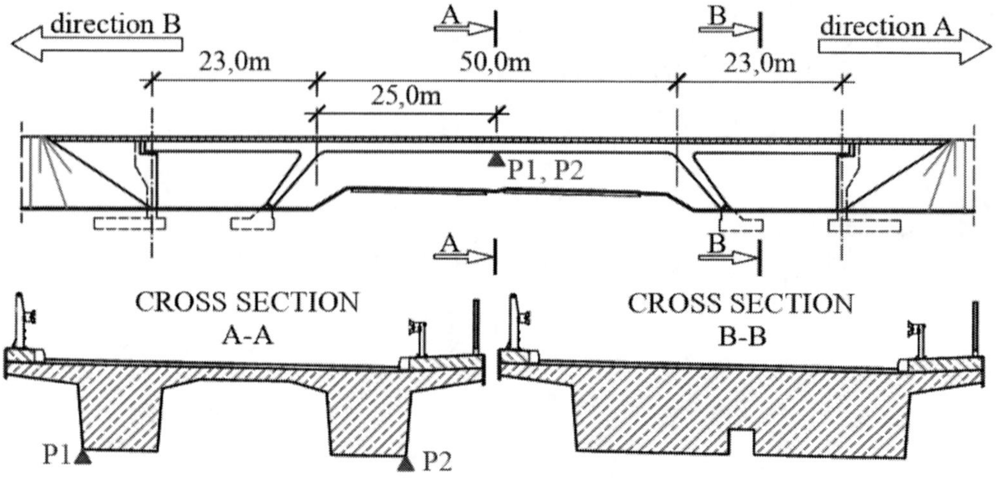

Fig. 5. The construction scheme of the W-2 viaduct with indicated measurement points.

For the damping analysis of the W-2 viaduct, ten signals were selected. Despite the high excitation, the signals which were acquired during the passage over the threshold located in the middle span had to be rejected. Before the last axle of loading vehicle left the structure, the vibrations dissipated (fig. 6a). As anticipated, the appropriate free vibration signal was recorded during the passage over the threshold located in the side span (last span in driving direction) – fig. 6b. The example of LD analysis for the W-2 viaduct is presented in the figures 7a and 7b. The results of calculations for the fundamental frequency f=2.83 Hz are shown in the table 2.

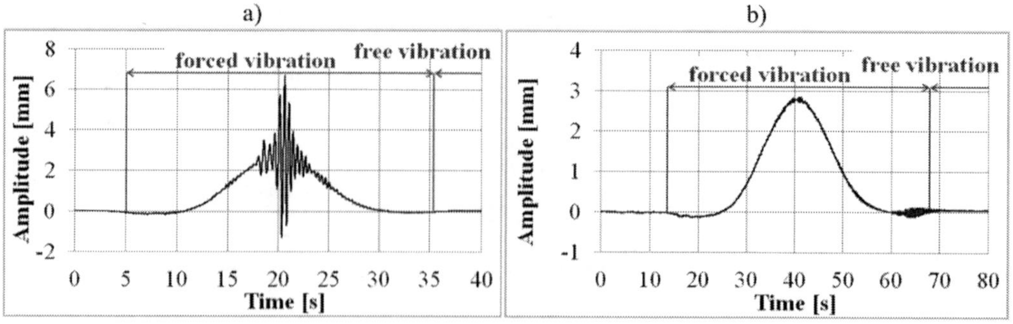

Fig. 6. The signal analysis for passages with the threshold located on the road: a) threshold in the middle span, b) threshold in the side span.

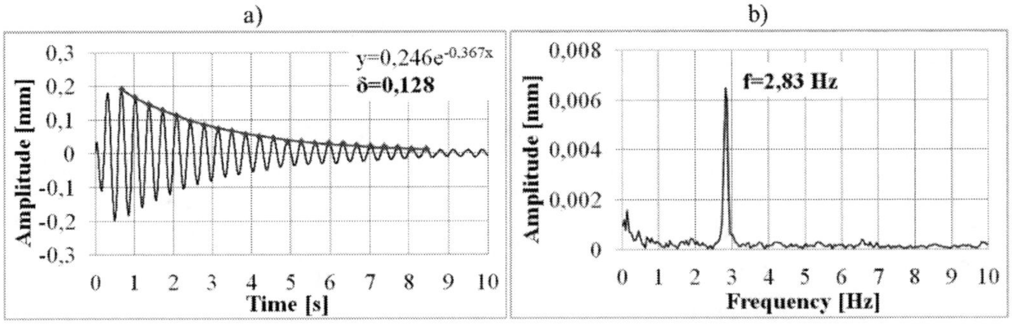

Fig. 7. The vibration damping analysis for a passage at a velocity of 40 km/h in the A direction: a) free vibrations and exponential curve fit to the vibration amplitudes, b) vibration spectrum.

Table 2. The values of LDT and damping ratio for the W-2 viaduct.

Velocity/direction of passage	Measurement point	Initial amplitude [mm]	LD	Damping ratio
30/B	P1	0,010	u.c.	u.c.
	P2	0,009	u.c.	u.c.
40/A (passage 1)	P1	0,191	0,128	0,021
	P2	0,185	0,132	0,022
40/A (passage 2)	P1	0,146	0,137	0,020
	P2	0,152	0,147	0,021
50/B (passage 1)	P1	0,128	0,141	0,022
	P2	0,123	0,154	0,023
50/B (passage 2)	P1	0,162	0,133	0,022
	P2	0,171	0,136	0,025
5/A (threshold in the side span)	P1	0,038	0,135	0,021
	P2	0,038	0,140	0,022
10/A (threshold in the middle span)	P1	0,013	u.c.	u.c.
	P2	0,014	u.c.	u.c.
		Mean value:	**0,138**	**0,022**
		Standard deviation:	0,007	0,001

u.c. – unable to calculate LD due to the low excitation of the construction

5 Summary and conclusions

The results presented in the paper indicate the damping level of two viaducts, which substructures are made of the post-tensioned concrete. The first viaduct is a typical modern two-span structure consisting of two trapezium girders in the static scheme of continuous beam. Second viaduct is a frame structure.

The values of logarithmic decrement and damping ratio were calculated using the curve-fitting approach adopting the least-squares method. The analysis was carried out basing on the displacement signals recorded during dynamic load tests of the viaducts. The signals with amplitudes lower than accuracy of the measurement equipment were rejected from the calculations.

The analysis indicates that not all signals recorded during field tests can be used to determine the level of damping – even if the excitation of forced vibration is high. In some cases, higher excitation of free vibration was noted during passages without the threshold located on the road. Vibration induced by the vehicle driving over the threshold often declines before last axle of the truck leaves the structure. It is caused by the low passage velocity. This problem occurs in the results of the W-2 viaduct dynamic load test (fig. 6a, table 2).

The determined values of LD varies from 0.82 to 0.99 for viaduct W-1 and from 0.128 to 0.154 for W-2. Differences in the estimation of damping for individual bridges result from the non-linearity of the phenomenon - e.g. the level of damping depends on the level of vibration. The mean values of LD for the analysed viaducts amount 0.088 for W-1 and 0.138 for W-2. The damping level of both constructions is acceptable – the viaducts can by assigned to an average damping group.

References

[1] CEN, *EN 1991-2 – Eurocode 1: Actions on structures, part 2: traffic loads on bridges*, European Committee for Standardisation, Brussels, (2003)

[2] M. Schlaich, K. Brownlie, J. Conzett et al,. fib Bulletin 32: Guidelines for the design of footbridges (Stuttgart: Sprint - Digital – Druck and fib) FIB, (2005)

[3] M. Pańtak, B. Jarek, K. Marecik, Vibration damping in steel footbridges, *IOP Conf. Series: Materials Science and Engineering 419*, (2018)

[4] T. Siwowski, L. Janas: Numerical and experimental assessment of the dynamic behaviour of a circular footbridge, *Proceedings of the International Conference on Structural Dynamic, EURODYN,2014-January*, 1297-1302, (2014)

[5] P. Hawryszków, R. Pimentel, F. Silva, Vibration effects of loads due to groups crossing a lively footbridge, Procedia Engineering, **199**, 2808-2813, (2017)

[6] G. Poprawa, M. Salamak, P. Klikowicz, Dynamic identification of composite beam pedestrian bridges, *Advances and Trends in Engineering Sciences and Technologies II - Proceedings of the 2nd International Conference on Engineering Sciences and Technologies, ESaT 2016*, 231-236, (2017)

[7] P. Olaszek, *Cyfrowe metody pomiarowe w zastosowaniu do badań mostów,* Komitet Inzynierii Ladowej i Wodnej PAN, Warsaw, (2015)

[8] ISO 18649 Mechanical vibration – Evaluation of measurement results from dynamic tests and investigations on bridges, (2004)

[9] P. Olaszek, L. Janas, M. Salamak, A. Matysek, Badania porównawcze przyspieszeń i przemieszczeń konstrukcji mostowych pod próbnym obciążeniem dynamicznym, Drogownictwo, **10**, 339-343, (2014)

[10] P. Olaszek, M. Łagoda, J. R. Casas, Diagnostic load testing and assessment of existing bridges: Examples of application, Structure and Infrastructure Engineering, **10**, 834–842, (2014)

[11] M. Salamak, *Experimental methods for determining the level of vibration damping in footbridges*, Silesian University of Technology, Faculty of Civil Engineering, (2003)

[12] PN-85/S-10030 Obiekty mostowe – obciążenia, (1985)

Stability of a hydraulic telescopic cylinder subjected to generalised load by a force directed towards the positive pole

Sebastian Uzny[1], and *Łukasz* Kutrowski[1,*]

[1]Institute of Mechanics and Machine Design Fundamentals, Czestochowa University of Technology, Dabrowskiego 73, 42-201 Czestochowa, Poland

Abstract. The paper presents the boundary problem of the stability of a telescopic hydraulic cylinder subjected to a generalized load with a force directed to the positive pole. The boundary problem was formulated on the basis of the Hamilton principle. Numerical calculations were carried out, taking into account the influence of the parameters of the load heads (radii of loading and receiving head, length of bolt). On the basis of the numerical calculations, regions of load heads parameters were presented, at which the load bearing capacity of the analysed telescopic hydraulic cylinder is the largest from the buckling standpoint.

1 Introduction

Hydraulic cylinders are certain types of engines in which the energy of the hydraulic fluid is converted into the mechanical energy of the translational movement. Apart from hydraulic systems, pneumatic and electric actuators are also used. Hydraulic cylinders are characterized by a greater working force. For this reason, they are successfully used in drives of individual members of construction machines, braking systems or presses with high pressure forces. Hydraulic cylinders can be classified due to many criteria. Due to the number of cylinder stages, one-stage and multistage cylinders can be distinguished. Multistage cylinders are also called telescopic cylinders.

As a result of the use of a larger number of cylinder members, an exceptionally long output travel from a very compact retracted length can be achieved. Hydraulic telescopic cylinders are mainly used in lifting systems of semi-trailers of trucks, drives of working members in basket lifts and elevators with hydraulic drives.

Hydraulic cylinders can be damaged due to many reasons. One of the common failures is the failure of hydraulic seals associated with damage to the cylinders or the piston rod of the actuator. When working in various types of constructions, the actuators can be also exposed to adverse vibrations, which can significantly reduce the service life of the cylinders. Due to the large slenderness of the cylinders, and especially telescopic structures, the basic problem during their design is the buckling phenomenon.

[*] Corresponding author: kutrowski@imipkm.pcz.pl

The stability analysis of single-stage hydraulic cylinders were considered in many publications, including [1-8]. Paper [1] presents a mathematical and physical model of a mining support, the construction of which corresponds to a single-stage hydraulic actuator. The results of numerical calculations and experimental tests related to the stability of a single-stage hydraulic cylinder are presented in [2]. Paper [3] presents a theoretical model of a single-stage hydraulic cylinder, taking into account a number of parameters affecting its stability. In publication [4] the effect of inaccuracies between the members of a single-stage cylinder on its stability were considered. The results of numerical and experimental tests were presented. The stability and effort of cylinder material of a single-stage hydraulic cylinder was the subject of work [5]. The flexural rigidity asymmetry factor was defined, and the regions of this parameter were presented with reference to the cylinder and piston rod of the actuator, in which the system is damaged as a result of buckling and effort of the cylinder material. The influence of stiffness increase in the supports of the single-stage actuator on its stability is presented in the publication [6]. The stiffness model of the hydraulic cylinder, which takes into account: axial piston rod deformation, cylinder volumetric expansion, finite stiffness of the supply lines, elasticity of the hydraulic medium, presence of air bubbles in the hydraulic fluid and deformability of the sealing elements are presented in [7]. The comparison of the results of numerical calculations obtained using FEM and the recommendations contained in DNV GL were presented in work [8].

The stability of the hydraulic telescopic cylinder subjected to Euler's load was considered in works [9, 10]. The publication [9] presents the influence of the elastic connection between the following cylinder elements on its stability. In the calculations Lame's theory for thick pipes was taken into account and the ranges of the cylinder barrel thickness parameter were determined, in which the destruction of the system is caused by the loss of stability and the effort of the cylinder material. Work [10] covers the effort analysis of a telescopic hydraulic cylinder. For a series of permissible stresses of the cylinder material, the influence of the cylinder barrel thickness parameter and the seal elements thickness parameter on the strength of individual cylinder barrels is presented. In correlation with buckling curves, the parameter ranges are defined in which the system is destroyed as a result of buckling or as a result of the effort of the material – which cause destruction within the weakest member of the actuator.

Generalized load with force directed to the positive pole was formulated by Tomski and belongs to the group of specific loads [11-14]. This load is caused by external force, directed to the fixed point (pole), which is located on the undeformed axis of the system. In addition, this pole is below the loaded end of the system (cf. fig. 1). The implementation of this load can be carried out in two ways: with the use of linear elements or circular elements. In this work, the case of this load is considered via circular elements.

2 Boundary problem

The considered system - a telescopic hydraulic cylinder subjected to a compressive generalized load with a force directed to the positive pole is shown in fig 1. This system consists of n elements, where elements from 1 to (n-1) are cylinders, and element n is the piston rod. In this system, it is assumed that the stiffness between following cylinder members is infinite. Each cylinder member is characterized by flexural rigidity $(EI)_i$, mass per length $(\rho A)_i$. In addition, fluid mass per length $(\rho A)_{ci}$ which fills the cylinders is taken into account (in the case of piston rod $(\rho A)_{cn} = 0$). The dimensions of each of the cylinder members depend on the diameter of the piston rod and the thickness of the cylinder barrels and sealing elements. In the work, subsequent diameters of the members are calculated using the following formulas:

$$d_{zi} = d_t + 2(n-i)g_U + 2(n-i)g_R \tag{1}$$

$$d_{wi} = d_t + 2(n-i)g_U + 2(n-1)g_R \tag{2}$$

where g_U i g_R are thicknesses of seals and cylinder barrels.

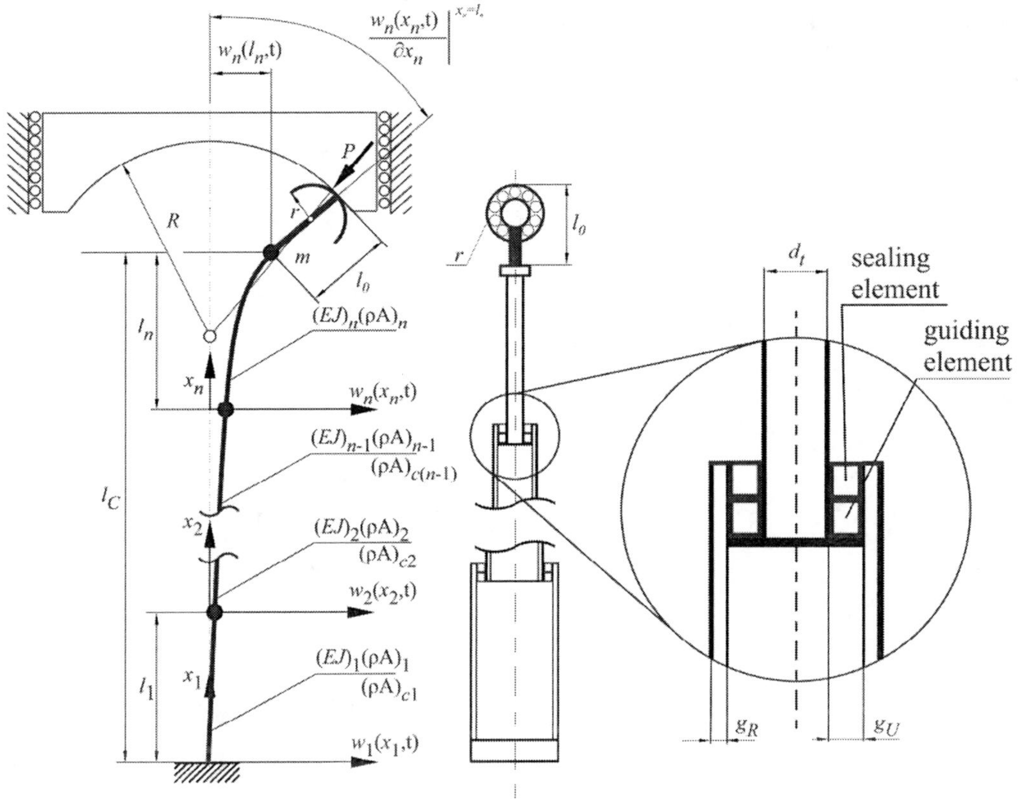

Fig. 1. Scheme of a hydraulic telescopic cylinder subjected to generalized load by a force directed towards the positive pole.

In this work, the critical load of the system is determined using the kinetic stability criterion. The boundary problem of free vibrations was formulated based on the Hamilton principle, taking into account the model of transverse vibrations of the cylinder proposed by Tomski [1]:

$$\delta \int_{t1}^{t2} (T-V)dt = 0 \tag{3}$$

Potential energy V and kinetic energy T can be written as follows:

$$V = \frac{1}{2}\sum_{i=1}^{n}\int_{0}^{l_i}(EJ)_i\left(\frac{\partial^2 w_i(x_i,t)}{\partial x_i^2}\right)^2 dx_i - \frac{1}{2}P\sum_{i=1}^{n}\int_{0}^{l_i}\left(\frac{\partial w_i(x_i,t)}{\partial x_i}\right)^2 dx_i +$$
$$+\frac{1}{2}\frac{P(l_0-r)}{R-r}\left(\frac{w_n(l_n,t)}{l_0-r}+\frac{\partial w_n(x_n,t)}{\partial x_n}\bigg|^{x_n=l_n}+w_n(l_n,t)+\frac{\partial w_n(x_n,t)}{\partial x_n}\bigg|^{x_n=l_n}\frac{(r-l_0)(R-r)}{l_0-r}\right) \tag{4}$$

$$T=\frac{1}{2}\sum_{i=1}^{n}(\rho A)_i\int_0^{l_i}\left(\frac{\partial w_i(x_i,t)}{\partial t}\right)^2 dx_i+\frac{1}{2}\sum_{i=1}^{n-1}(\rho A)_{ci}\int_0^{l_i}\left(\frac{\partial w_i(x_i,t)}{\partial t}\right)^2 dx_i+ \\ +\frac{1}{2}m\left(\left.\frac{\partial w_n(x_n,t)}{\partial t}\right|^{x_n=l_n}\right)^2 \tag{5}$$

After appropriate mathematical transformations, equations of motion and natural boundary conditions can be obtained. After separation of variables, the equations can be written in the form:

$$(EI)_i\frac{d^4y_i(x_i)}{dx_i^4}+P\frac{d^2y_i(x_i)}{dx_i^2}-(\rho A)_i\,\omega^2\frac{d^2y_i(x_i)}{dx_i^2}+ \\ -(\rho A)_{ci}\,\omega^2\frac{d^2y_i(x_i)}{dx_i^2}=0 \qquad for\ i=1,2...n \tag{6}$$

$$(EI)_i\left.\frac{\partial^3y_i(x_i)}{\partial x_i^3}\right|^{x_i=l_i}-(EI)_{i+1}\left.\frac{\partial^3y_{i+1}(x_{i+1})}{\partial x_{i+1}^3}\right|_{x_{i+1}=0}+P\left.\frac{\partial y_i(x_i)}{\partial x_i}\right|^{x_i=l_i}-P\left.\frac{\partial y_{i+1}(x_{i+1})}{\partial x_{i+1}}\right|_{x_{i+1}=0}=0 \tag{7}$$

$$-(EI)_i\left.\frac{\partial^2y_i(x_i)}{\partial x_i^2}\right|^{x_i=l_i}+(EI)_{i+1}\left.\frac{\partial^2y_{i+1}(x_{i+1})}{\partial x_{i+1}^2}\right|_{x_{i+1}=0}=0 \tag{8}$$

$$(EI)_n\left.\frac{\partial^3y_n(x_n)}{\partial x_n^3}\right|^{x_n=l_n}+P\left(\frac{r-l_0}{R-r}\right)\left(\left.\frac{\partial y_n(x_n)}{\partial x_n}\right|^{x_n=l_n}(R-l_0)-y_n(l_n)\right)=0 \tag{9}$$

$$(EI)_n\left.\frac{\partial^3y_n(x_n)}{\partial x_n^3}\right|^{x_n=l_n}+P\left(\frac{1}{R-r}\right)\left(\left.\frac{\partial y_n(x_n)}{\partial x_n}\right|^{x_n=l_n}(R-l_0)-y_n(l_n)\right)+m\omega^2y_n(l_n)=0 \tag{10}$$

In the analysed system, the geometrical boundary conditions are distinguished:

$$y_i(l_i)=y_{i+1}(0) \tag{11}$$

$$\left.\frac{\partial y_i(x_i)}{\partial x_i}\right|^{x_i=l_i}=\left.\frac{\partial y_{i+1}(x_{i+1})}{\partial x_{i+1}}\right|_{x_{i+1}=0} \tag{12}$$

$$y_1(0)=0 \quad ; \quad \left.\frac{dy_1(x_1)}{dx_1}\right|_{x=0}=0 \tag{13, 14}$$

The solution of equations of motion is presented in form:

$$y_i(x_i)=A_i\cosh(\alpha_ix_i)+B_i\sinh(\alpha_ix_i)+C_i\cos(\beta_ix_i)+D_i\sin(\beta_ix_i) \tag{15}$$

where:

$$\alpha_i=\sqrt{-\frac{k_i^2}{2}+\sqrt{\frac{k_i^4}{4}+\Omega_i^2}} \ ;\ \beta_i=\sqrt{\frac{k_i^2}{2}+\sqrt{\frac{k_i^4}{4}+\Omega_i^2}} \tag{16a,b}$$

$$\Omega_i^2=\frac{((\rho A)_i+(\rho A)_{ci})\omega^2}{(EI)_i} \ ;\ k_i^2=\frac{P}{(EI)_i} \tag{17a,b}$$

Taking into account the equations (6-14) in the solution (15), the system of equations is obtained. The matrix determinant of coefficients equated to zero leads to the transcendental equation, from which the natural vibration frequency can be obtained. The critical load of the analysed system corresponds to the first natural frequency equal to zero.

3 Results of numerical simulations

Results of numerical calculations are presented by using non-dimensional parameters:

$$\zeta_{GU} = \frac{g_U}{d_t};\ \zeta_{GR} = \frac{g_R}{d_t};\ \zeta_L = \frac{l_C}{d_t};\ \lambda_{cr} = \frac{P_{cr} l_C^2}{(EI)_n};\ \zeta_{l0} = \frac{l_0}{R};\ \zeta_R = \frac{R}{l_C};\ \zeta_r = \frac{r}{R}$$

where:

ζ_{GU} – parameter of sealing and guiding elements thickness
ζ_{GR} – parameter of cylinder barrel thickness
ζ_L – parameter of cylinder length
λ_{cr} – parameter of critical load
ζ_{l0} – parameter of bolt length
ζ_R – parameter of loading head radius
ζ_r – parameter of receiving head radius

Fig. 2-5 shows the effect of three dimensionless parameters of load heads (ζ_{l0}, ζ_R, ζ_r) on the stability of a telescopic hydraulic cylinder with three and five stages.

Based on the results presented in fig. 2-3, it was noticed that the ratio of the radius of the receiving head to the radius of the head inducing the load of the system (the relationship represented by parameter ζ_r) has a significant impact on the critical load. In the range of this parameter ⟨0.6; 0.9⟩, the critical load increases when the ζ_R parameter is increased (at lower values). At higher values of the parameter ζ_R the critical load of the system decreases with increasing this parameter. In the case when the parameter ζ_r is lower than or equal to 0.5 - the increase of the radius of the loading head (increase of the parameter ζ_R) causes a significant decrease of the load capacity of the system. The trend of curves changes $\lambda_{cr}(\zeta_R)$ is the same for a three and five stages cylinder.

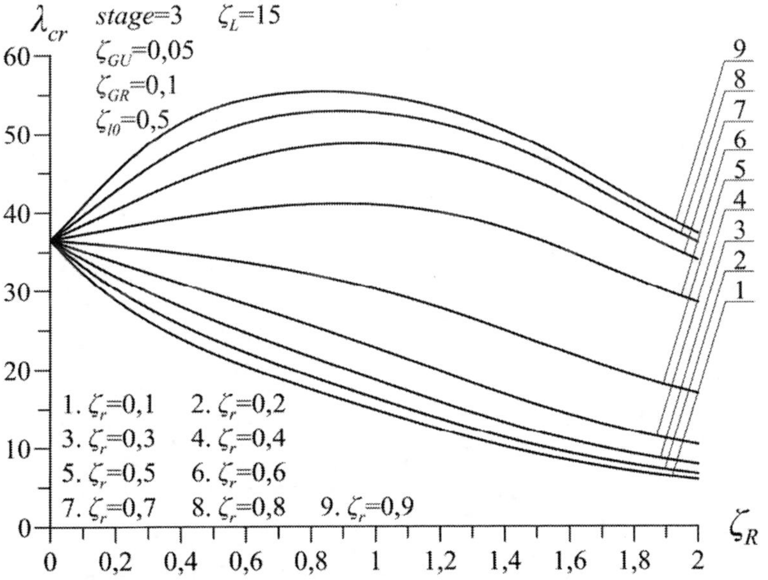

Fig. 2. The influence of parameter ζ_R (for different parameters ζ_r) on stability of three stages hydraulic cylinder.

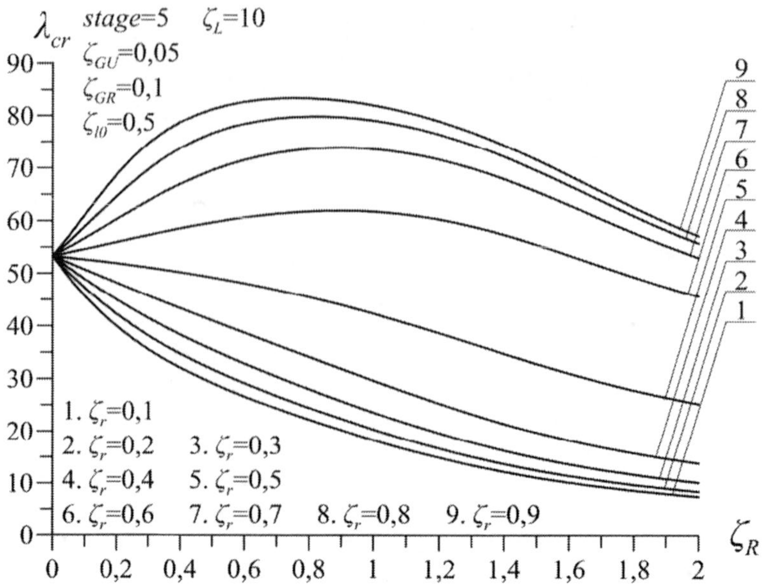

Fig. 3. The influence of parameter ζ_R (for different parameters ζ_r) on stability of five stages hydraulic cylinder.

The influence of the bolt length on the structural stability is shown in fig. 4-5. It was shown that with the assumed system parameters in the case when the parameter ζ_{l0} is from the interval ⟨0; 0,6⟩, the critical load of the system is increasing when the radius of the loading head is increased in a certain range. The greater the length of the bolt, the higher the critical load increase is. Increasing the length of the bolt (from the range $\zeta_{l0} \in \langle 0;\ 0{,}6 \rangle$) the maximum critical load occurs at higher radius of the loading head. Increasing the length of the bolt while increasing the radii of the load heads causes a decrease in the critical load.

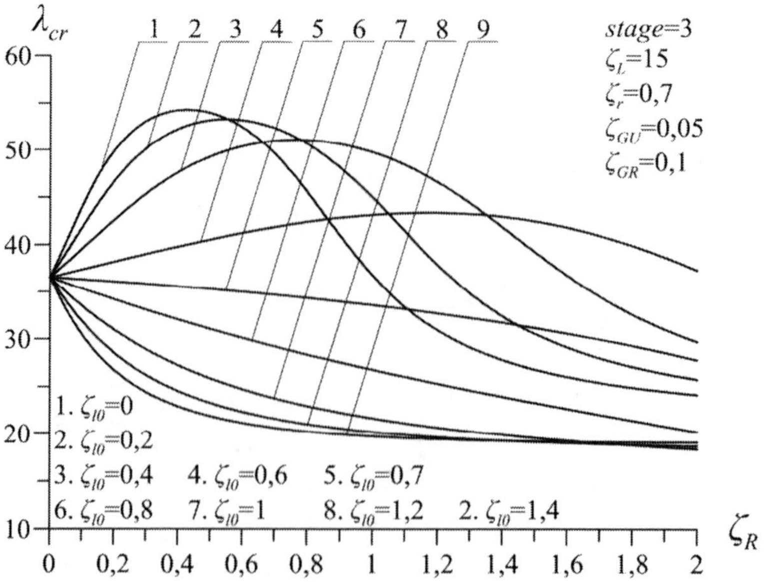

Fig. 4. The influence of parameter ζ_R (for different parameters ζ_{l0}) on stability of three stages hydraulic cylinder.

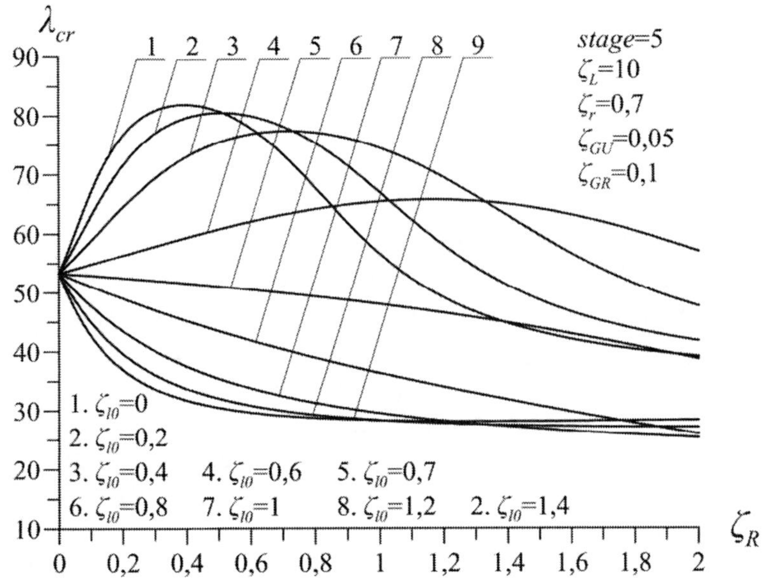

Fig. 5. The influence of parameter ζ_R (for different parameters ζ_{l0}) on stability of five stages hydraulic cylinder.

3 Conclusions

This paper presents the stability problem of a telescopic hydraulic cylinder subjected to a generalized load with a force directed towards to the positive pole. The boundary problem was formulated using the kinetic criterion of stability. The equations of motion and boundary conditions of the considered system are presented. With specific assumptions regarding the thickness of cylinders and seals, numerical simulations were carried out regarding the critical load of the system depending on the parameters of the load heads. Based on the obtained critical load distributions, the intervals of the parameters of the load heads, at which the considered system is characterized by the highest load capacity, can be determined. As a continuation of the research, the impact of the parameters of the load heads implementing the load considered in this work should be determined on the characteristic curves determined on the plane of load - natural frequency.

The study has been carried out within the statutory funds of the Czestochowa University of Technology (BS/PB 1-101-3020/17/P).

References

1. L. Tomski, Engineering Transactions. **25**(2). 247-263 (1977)
2. L. Tomski, S. Uzny, Zeszyty Naukowe Politechniki Rzeszowskiej. seria: Mechanika. **258**(74). 369-380 (2008)
3. V. Gómez Rodríguez, J. Jose Cabello Eras, H. Hernández Herrera, R. Goytisolo, IJET. **8**(6). 2443-2451 (2016/2017)
4. P.J. Gamez-Montero, E. Salazar, R. Castilla, J. Freire, M. Khamashta, E. Codina, IJMS, **51**(2), 105-113 (2009)

5. L. Tomski, S. Uzny, ACME, **11**(3), 769-785 (2011)
6. S. Uzny, Proc. Appl. Math. Mech., **9**, 303-304 (2009)
7. F. Hao, D. Qungui, H. Yuxian, Ch. Yongbin, STROJ VESTN-J MECH E., **63**(7-8), 447-456 (2017)
8. E. Narvydas, Mechanika, **22**(6), 474-477 (2016)
9. S. Uzny, K. Sokół, Ł. Kutrowski, In: Rusiński E., Pietrusiak D. (eds). *Proceedings of the 13th International Scientific Conference. Lecture Notes in Mechanical Engineering.* Springer, 581-588 (2017)
10. S. Uzny, Ł. Kutrowski, Modelowanie Inżynierskie, **68** , 125-131 (2019)
11. L. Tomski, *Drgania i stateczność układów smukłych*, Warszawa: WNT (2004)
12. L. Tomski, *Drgania swobodne i stateczność obiektów smukłych jako układów liniowych lub nieliniowych*, Warszawa: WNT (2007)
13. L. Tomski, *Drgania swobodne i stateczność układów smukłych poddanych obciążeniu konserwatywnemu lub niekonserwatywnemu*, Warszawa: PWN (2012)
14. L. Tomski, J. Szmidla, J THEOR APP MECH-POL, **42**(1), 163-193 (2004)

Dynamic tests of composite footbridge segment – experimental and numerical studies

Tomasz Wiczenbach[1], *Tomasz* Ferenc[1*], *Łukasz* Pyrzowski[1], and *Jacek* Chróścielewski[1]

[1]Faculty of Civil and Environmental Engineering, Gdansk University of Technology, Poland

Abstract. Experimental dynamic tests and numerical simulations of a 3 meters long segment of a footbridge made of polymer composites are presented in the paper. The span-length is reduced, however dimensions of cross-section are the same as the target footbridge. The segment structure is made of sandwich panels, which consist of fibre reinforced polymer (GFRP) laminates (skins) and a PET foam (core). The first part of the paper contains description and results of experimental dynamic tests, which were subjected to the structure. The excitations during these tests was realised by a group of people standing or jumping on the structure deck. The obtained results allowed to determine natural frequencies as well as damping ratios. Moreover, cyclic loading was applied to the structure deck to check the possibility of delamination or debonding occurrence, as a result of repetitive dynamic excitations. The second part contains validation of numerical model - results of numerical analysis and its comparison with the experimental ones.

1 Introduction

Nowadays, requirements for modern structure often force the use of new materials. In order to design structure made from untraditional material, like composites studied in the paper, it is obligatory to make more sophisticated analysis compared to the one made from traditional ones like steel or concrete. Material characteristics of composites are not standardised and often depend on many factors like components used in production as well as manufacturing process itself. First of all it is necessary to conduct experimental tests that allow to identify material parameters. Moreover, it is also obligatory to check correctness of assumptions made during numerical analysis by conducting validation test [1-2].

In the paper the investigation of dynamic behaviour of the structure which is thought to be a 3 metres long segment of GFRP composite footbridge is presented (Fig. 1a). The dynamic analysis is crucial for very light footbridges [3]. Their small mass results low natural frequencies that are close to the critical frequencies of dynamic excitations produced by pedestrians. The presented research was a starting point for the analysis of the target full-length footbridge (Fig. 1b). The study is a part of project FOBRIDGE realized in years 2013-2015 by consortium of Gdansk University of Technology (main leader), Military University of Technology in Warsaw and the private company ROMA Co. Ltd.

* Corresponding author: tomasz.ferenc@pg.edu.pl

The main aim of the study, was to construct a single span footbridge made only of composite materials as one element in a single technological process [4-6].

Fig. 1. Side views: (a) analysed segment, (b) target footbridge.

2 Description of segment

The structure of analysed segment (Fig. 1a) is a sandwich type. Its GFRP laminate skins are made of vinyl ester resin reinforced with six layers of two-directional glass fabrics. The thickness of a single layer is 0.663 mm, hence the thickness of the entire laminate is 3.978 mm. The density was assumed as 1.71 g/cm^3. The core is made of a PET foam. The thickness of the foam is 100 mm, and the density is 100 kg/m^3. Material parameters are listed in table 1 for single lamina [7] and in table 2 for PET foam, respectively.

Table 1. Material parameters for lamina [7].

Parameter		Description	value	unit
E_1	E_2	longitudinal (1) and transverse (2) elastic moduli	23.4	[GPa]
ν_{12}		Poisson's ratio	0.153	[-]
G_{12}		in-plane shear modulus	3.52	[GPa]
G_{13}	G_{23}	transverse shear moduli	1.36	[GPa]

Table 2. Material parameters for PET foam.

Parameter	Description	value	unit
E	elastic modulus	70	[MPa]
ν_{12}	Poisson's ratio	0.4	[-]

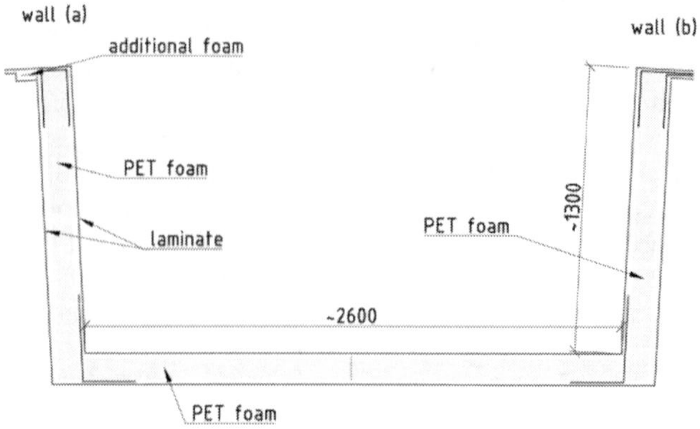

Fig. 2. Cross-section of segment [mm].

Segment has the same cross-section as the target footbridge (Fig. 2), its total length is reduced to 3 metres, thus theoretical length is 2.5 m. The usable width of segment is 2.59 m and the height of handrails is 1.3 m. The dimensions of cross-section are adopted in accordance with the polish standards. Additionally, segment is not symmetric, parapet in wall (a) is strengthened by an additional 10 mm PET foam. During tests the segment was situated on four rubber bearings with dimensions of 300×300×30 mm.

The main aim of segment behaviour study was to validate numerical model that would be used in further design process of a target footbridge with length of 14 meters. Moreover, manufacturing of segment itself was also a technological test. Segment, the same as the target footbridge, was produced as a single element in one cycle of the infusion process.

3 Conducted experiments

To obtain measured values the following instrumentation have been installed: strain gauges (T) and displacement sensors (U). Location of this sensors is shown in Fig. 3. Additionally, the lines indicate the direction of the measured values.

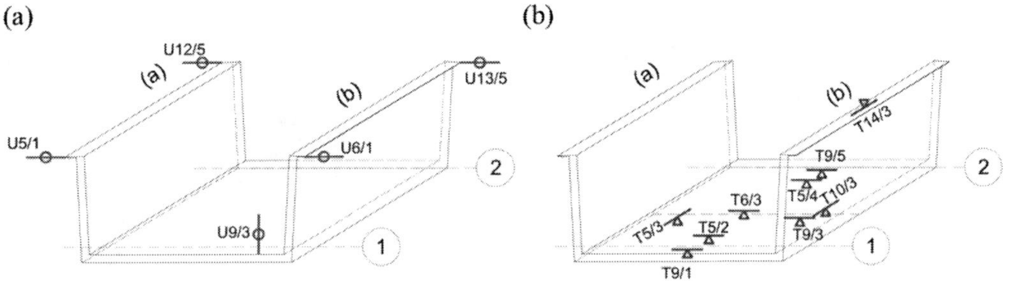

Fig. 3. Location of devices: (a) displacements sensors U, (b) strain gauges T.

3.1 Dynamic loading – group of six people

The segment was subjected by dynamic loading. For static test examination see [8-9]. Scheme S1 (Fig. 4a) consisted of forces applied on the platform slab realized by a group of six people with a total weight of 4.8 kN. The course of the test can be observed on the graph presenting vertical displacement at the midpoint of the platform (point U9/3) in a function of time (Fig. 4b).

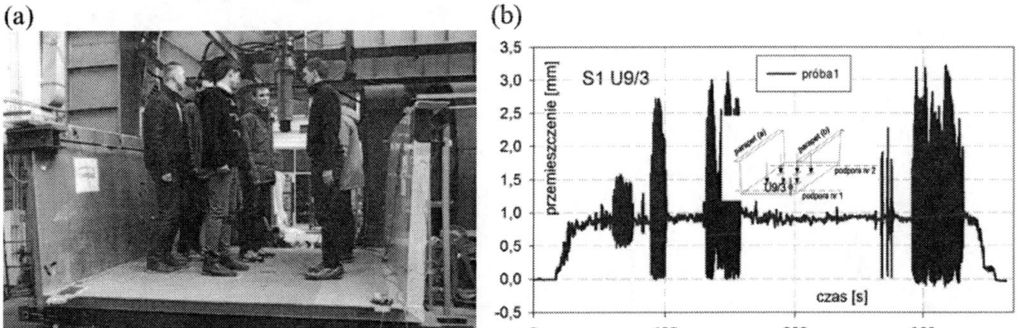

Fig. 4. Load test S1: (a) excitation, (b) displacement in point U9/3.

Conducted experiments allowed to determine the dynamic characteristics of the segment. Using the fast Fourier transform (FFT), it was possible to determine the natural frequencies. Based on the displacement time history registered at point U5/1 (Fig. 5a) and U12/1 (Fig. 6a) for wall (a) and U6/1 (Fig. 7a) and U13/5 (Fig. 8a) for wall (b), the Fourier transform was conducted (Figs. 5b-8b).

(a) (b)

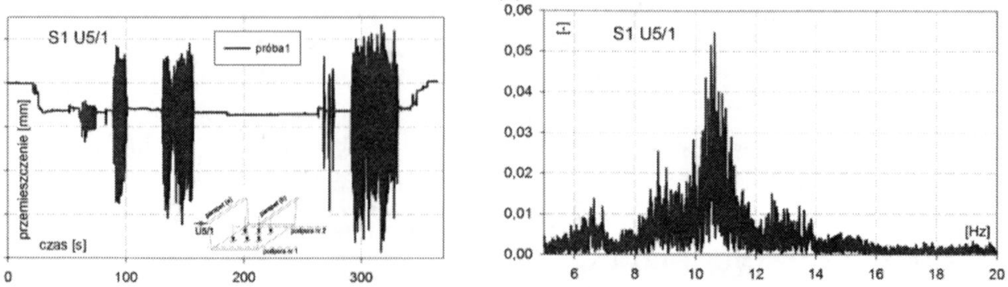

Fig. 5. Scheme S1, point U5/1: (a) acceleration-time relation, (b) Fourier transform.

(a) (b)

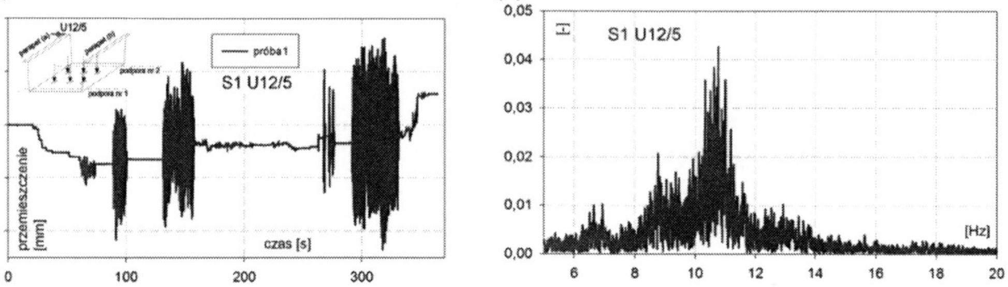

Fig. 6. Scheme S1, point U12/5: (a) acceleration-time relation, (b) Fourier transform.

(a) (b)

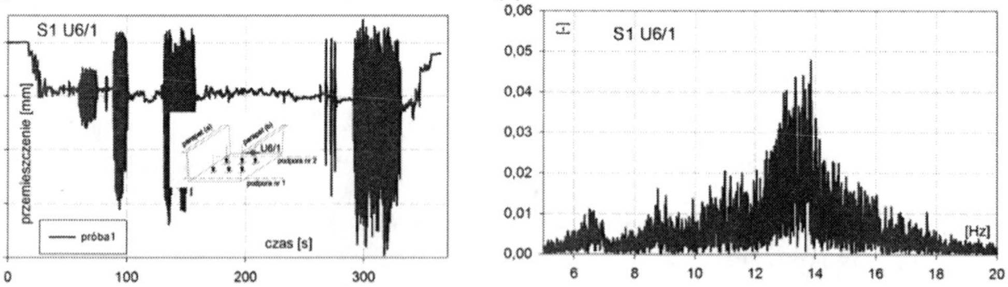

Fig. 7. Scheme S1, point U6/1: (a) acceleration-time relation, (b) Fourier transform.

(a) (b)

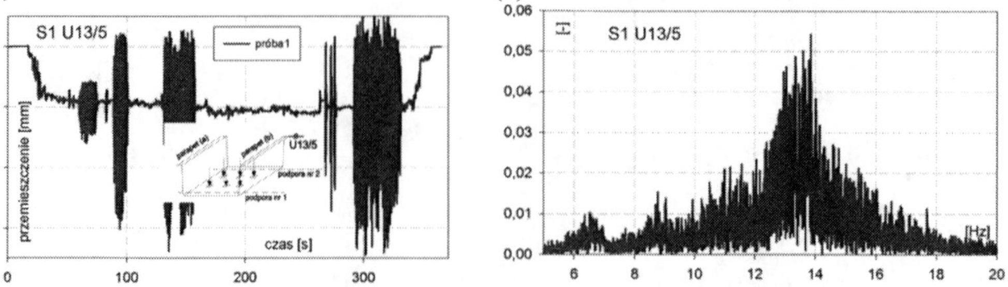

Fig. 8. Scheme S1, point U13/5: (a) acceleration-time relation, (b) Fourier transform.

The obtained values of natural frequencies are presented in table 3. Two different values of the natural vibration frequencies result from non-symmetrical structure of the segment (parapet strengthening on one side).

Table 3. Scheme S1: Natural frequencies.

Point	Localization	Frequency [Hz]
U5/1	wall (a)	10.6
U12/5	wall (a)	10.7
U6/1	wall (b)	13.9
U13/5	wall (b)	13.9

Based on the research, by measuring the amplitude of vibrations in subsequent cycles, it was possible to estimate logarithmic damping decrement δ and hence damping ratio ξ, in accordance with the formula (for small values of damping ratio):

$$\delta = \frac{1}{j} ln \frac{u_i}{u_{i+j}} = \frac{2\pi\xi}{\sqrt{1-\xi^2}} \rightarrow \xi \cong \frac{\delta}{2\pi}, \tag{1}$$

where j – no. of cycles, u_i – amplitude of vibration in the i-th cycle, u_{i+j} – the vibration amplitude in $(i+j)$-th cycle.

The values of the damping ratios that were obtained by analysing the displacements at the points U5/1, U6/1, U12/5 and U13/5 is summarized in table 4.

Table 4. Logarithmic decrement of damping and damping ratio.

Point	Logarithmic decrement of damping	Damping ratio
U5/1	0.327	0.052
U12/5	0.188	0.03
U6/1	0.390	0.062
U13/5	0.320	0.051
av.	0.306	0.049

3.2 Dynamic loading – cyclic force

During the cyclic test the segment was subjected to a single force load with frequency of 1 Hz. The load was applied in the middle of the deck by a hydraulic piston, see Fig. 9.

(a) (b)

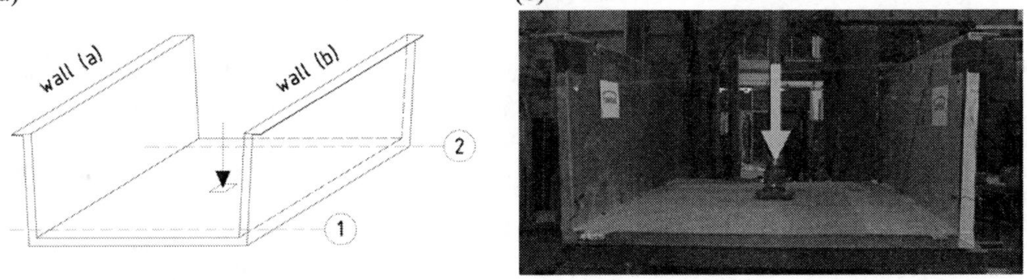

Fig. 9. Scheme of cyclic force application.

The main purpose of conducted experiment was to investigate if potentially delamination in laminate or debonding in sandwich structure occur. Hence, 30000 cycles of harmonic force was applied, with range from 12 kN to 50 kN. Displacements and strains in several points (Fig. 3) were measured. The examples of displacement time history from different ranges of the test in measuring point U9/3 are shown in Fig. 10. As it can be observed, the structural response caused by harmonic excitation also give harmonic function. The range of displacement is from about 4 mm to 14 mm and almost do not change through the experiment. No external and internal failures of the structure were observed during all the conducted tests, as well as later after cutting the segment into pieces.

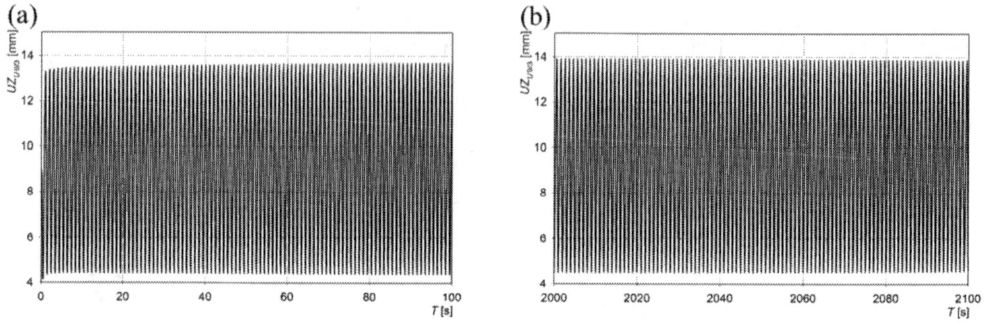

Fig. 10. Displacements in point U9/3: (a) 0-100 cycles range, (b) 2000-2100 cycles range.

4 Numerical simulations

The numerical model of the analysed segment was made in the FEMAP with Nastran software with using the finite element method FEM (Fig. 11).

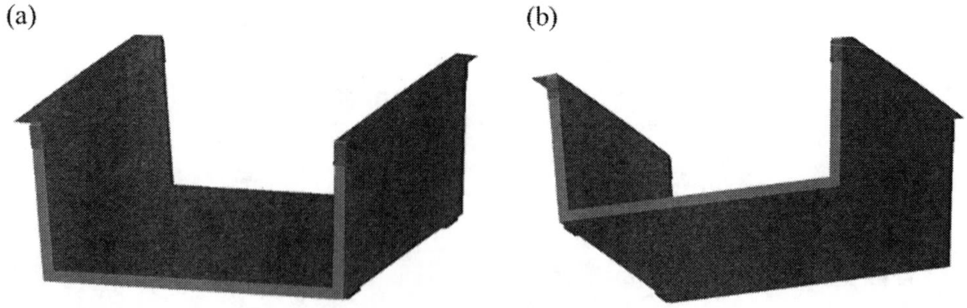

Fig. 11. Visualization of the FEM model: (a) general view, (b) general view from below.

Laminates were modelled as multi-layered shells while PET as solid elements. Moreover, laminates were modelled by means of equivalent single layer (ESL) theory, where 3D laminate is simplified by 2D shell. Material parameters were assumed according to tables 1-2 for single lamina and PET foam, respectively. Four-node element were used for shells with linear shape functions and full integration, while for solids eight-node element also with liner shape functions and full integration. The characteristic dimension of the finite element is assumed at the level of about 25 mm, which gives four elements through the thickness of the sandwich structure in the deck, and five elements in the walls.

https://doi.org/10.1051/matecconf/201928500021

Thus, the total number of finite elements in the model is 199 498, while the number of nodes is 157 401.

After model analysis was conducted, the first four mode shapes and natural frequencies was determined and shown in Fig. 12. They were determined in two approaches: the first one without considering mass of group of people standing on the deck and the second one taking into account that mass. The first frequency was obtained as 12.31 Hz without mass and 11.24 Hz with mass, the second as 16.34 Hz with and 13.60 Hz without mass, the third as 20.66 Hz with and 15.31 Hz without mass, and the fourth 30.18 Hz with and 17.94 Hz without mass. The results confirmed asymmetric structure response, Fig. 12a shows the first mode shape where wall (a) vibrates, while Fig. 12b shows vibrations of wall (b).

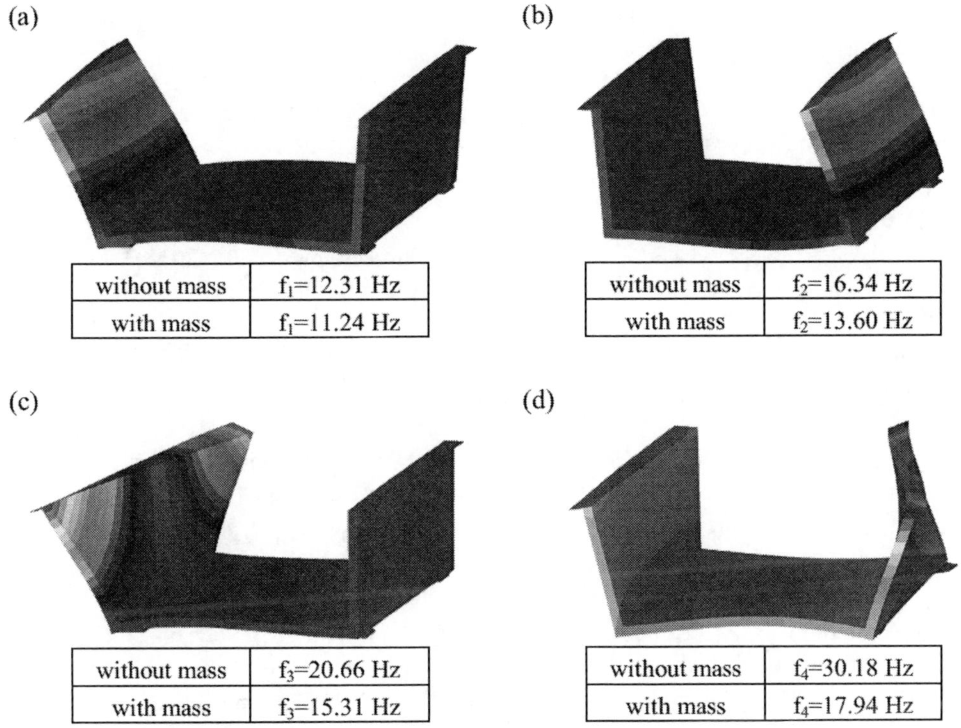

Fig. 12. Mode shapes and natural frequencies: (a) first, (b) second, (c) third, (d) fourth.

5 Conclusions

In the paper, authors focused on analysis of 3 meters long segment of the target footbridge made fully of composite materials. The segment was studied in an aspect of its dynamic behaviour. Parameters such as natural frequencies and damping ratios were determined.

The results presented in this paper were based on experimental tests supported by numerical simulations. Asymmetric cross-section can cause various frequencies of free vibrations of two different walls (a) and (b). The natural frequencies of segment obtained from experimental results were estimated as 10.7 Hz in the wall (a) and 13.9 Hz in the wall (b). Values for numerical models are various according to approach – whether mass is taken into account or not. The obtained values compared with experimental results are listed in table 5.

Table 5. Comparison of natural frequencies.

wall	experiment	numerical model	
		without mass	with mass
(a)	10.65	12.31	11.24
(b)	13.9	16.34	13.60

Some differences of values obtained in experimental and numerical part could be caused by few reasons. First of all, oscillations in experiment were excited by jumps of group of people. As it is shown, there is significant difference whether their mass is considered or not. More accurate results were obtained with mass. Furthermore, boundary conditions also can affect behaviour of structure. It was assumed, that segment is rigidly connected to rubber bearings. In fact, during vibrations segment could lose contact between itself and bearings.

As it was presented, developed modelling method of shell and sandwich type structure gave satisfying results – the model was positively validated (see also [8-9]) and after expansion was successfully used to design target composite footbridge [4-5].

References

1. Pyrzowski Ł., Sobczyk B., Witkowski W., Chróścielewski J. (2016) Three-point bending test of sandwich beams supporting the GFRP footbridge design process – validation. 3rd Polish Congress of Mechanics (PCM) / 21st International Conference on Computer Methods in Mechanics (CMM), Taylor & Francis Group, 489-492
2. Siwowski T., Kaleta D., Rajchel M. (2018) Structural behaviour of an all-composite road bridge. Composite Structures, **192**, 555–567
3. Drygala I.J., Polak M.A., Dulinska J.M. (2019) Vibration serviceability assessment of GFRP pedestrian bridges, Engineering Structures, **184**, 176-185
4. Chróścielewski J., Miśkiewicz M., Pyrzowski Ł., Sobczyk B., Wilde K. (2017) A novel sandwich footbridge - Practical application of laminated composites in bridge design and in situ measurements of static response, Composites Part B **126**, 153-161
5. Chróścielewski J., Miśkiewicz M., Pyrzowski Ł., Rucka M., Sobczyk B., Wilde K (2018) Modal properties identification of a novel sandwich footbridge – Comparison of measured dynamic response and FEA, Composites Part B **151**, 245-255
6. Ferenc T., Pyrzowski Ł., Chróścielewski J., Mikulski T. (2018) Sensitivity analysis in designing process of sandwich U-shaped composite footbridge, Shell Structures: Theory and Applications. - Vol. **4**, ed. W. Pietraszkiewicz, W. Witkowski, Leiden: CRC Press/Balkema, 413-416
7. Klasztorny M., Gotowicki P., Nycz D. 2013. Identification tests of new soaked composite BG/F in temperature 20°C, Report 6-2013v4 of project Fobridge, (in Polish)
8. Miśkiewicz M., Daszkiewicz K., Ferenc T., Witkowski W., Chróścielewski J. (2016) Experimental tests and numerical simulations of full scale composite sandwich segment of a foot- and cycle- bridge. Advances in Mechanics: Theoretical, Computational and Interdisciplinary Issues, 2016, Taylor & Francis Group, 401-404
9. Chróścielewski J., Ferenc T., Mikulski T., Miśkiewicz M., Pyrzowski Ł. (2019) Numerical modeling and experimental validation of full-scale segment to support design of novel GFRP footbridge, Composite Structures, **213**, 299-307

On estimation of occupant safety in vehicular crashes into roadside obstacles using non-linear dynamic analysis

Krzysztof Wilde[1], *Arkadiusz* Tilsen[1], *Stanisław* Burzyński[1,*], and *Wojciech* Witkowski[1]

[1]Gdansk University of Technology, Faculty of Civil and Environmental Engineering, Department of Mechanics of Materials, ul. Narutowicza 11/12, 80-233 Gdansk, Poland

Abstract. The article describes a comparison of two general methods of occupants safety estimation based on a numerical examples. The so-called direct method is mainly based on the HIC (Head Injury Criterion) of a crash test dummy in a vehicle with passive safety system while the indirect method uses a European standard approach to estimate impact severity level.

1 Introduction

The first automotive accident was registered in 1896 in Europe at speed equal to 6.4 km/h. Nowadays traffic accidents become a serious social problem. There are above 450 million drivers in Europe [1]. In 2017 there were 49 killed occupants per million citizens in EU.

Reduction of the number of fatal accidents is possible due to the use of active and passive safety car systems and road safety devices. Table 1 shows how various protection systems in frontal crashes influence the overall outcome of the accident.

Table 1. Reduction of fatal accidents due to protection systems [2].

Protection system	Reduction of fatal accidents
Airbag and seat-belt	50 %
Only airbag	14 %
Seat-belt	45 %

2 Methods

There are two general methods to evaluate occupant's safety. The indirect method corresponds to vehicle's dynamic response in traffic accident whereas the direct method is based on dummy that mimics the human) dynamic response.

* Corresponding author: staburzy@pg.edu.pl

2.1 Indirect method

The indirect method is the most popular in designing road restraint system due to its high efficiency and lowest costs. The method requires to analyse traffic accident using the vehicle and the obstacle. There is no possibility to evaluate passenger safety directly. According to European standard [4] occupant safety is estimated by ASI and THIV parameters. These values specify the impact severity level of the restraint system. Table 3 shows three level scale of impact severity. Level A affords a greater level of safety for the occupant of an errant car than level B, and level B greater than level C. Additionally standard [11] permits maximum values THIV = 44 km/h and ASI = 1.4 for high energy absorbing roadside equipment.

Table 3. Impact severity level for road barriers [4].

Impact severity level	Index values		
A	ASI ≤ 1.0	and	THIV ≤ 33 km/h
B	ASI ≤ 1.4		
C	ASI ≤ 1.9		

2.1.1 Acceleration Severity Index

The acceleration severity index (ASI) is a function of time, computed as follow [5]:

$$\mathrm{ASI}(t) = \left[\left(\frac{\bar{a}_x}{\hat{a}_x}\right)^2 + \left(\frac{\bar{a}_y}{\hat{a}_y}\right)^2 + \left(\frac{\bar{a}_z}{\hat{a}_z}\right)^2\right]^{0.5}, \tag{1}$$

where $\bar{a}_x$, $\bar{a}_y$ and $\bar{a}_z$ are the values of vehicle's center of gravity accelerations, filtered with a four-pole phaseless Butterworth low-pass digital filter with a cut-off frequency of 13 Hz. Values in denominator are the limit values of components which are equal: $\hat{a}_x = 12$ g, $\hat{a}_y = 9$ g and $\hat{a}_z = 10$ g (g denotes freefall acceleration). The maximum value of ASI is chosen to check the impact severity level.

2.1.2 Theoretical Head Impact Velocity

The theoretical head impact velocity (THIV) is represented by a point that hits the surface of the hypothetical vehicle interior. It is calculated according to procedure defined in European standard [5]. The time of flight of hypothetical head and value of THIV are calculated from horizontal velocities of the vehicle's canter of gravity and 1.2 x 0.6 m rectangle which represent the interior of the vehicle.

2.2 Direct method

Direct method is more complicated. It takes into account the occupant body and passive safety car systems like airbags and seatbelts in traffic accident. It is also more numerically and financially expensive. Different parts of the occupant's body have different injury mechanisms. This research is focused on head dynamic response. Abbreviated Injury Score (AIS) describes severity rating of occupant. Fatality range contains items such as injuries of organs, fractures and other injuries. Table 2 shows seven step scale of AIS [1].

Commonly used is the head injury criterion (HIC). Table 4 shows the relation between AIS code and HIC value. The table contains also probable injury's prognosis after accident.

Table 2. Abbreviated injury score (AIS) [1].

AIS code	Injury Level	Fatality Range
0	No injury	0.0 %
1	Minor	0.0 - 0.1 %
2	Moderate	0.1 - 0.4 %
3	Serious	0.8 - 2.1 %
4	Severe	7.9 - 10.6 %
5	Critical	53.1 - 58.4 %
6	Maximum	Virtually unsurvivable

Table 4. HIC, AIS code and head injury [6].

HIC	AIS	Head injury
135 - 519	1	Headache or dizziness
520 - 899	2	Unconscious less than 1 h; linear fracture
900 - 1254	3	Unconscious 1 – 6 h; depressed fracture
1255 - 1574	4	Unconscious 6 – 24 h; open fracture
1575 - 1859	5	Unconscious more than 24 h; large hematoma
>1860	6	Non-survivable

2.2.1 Head Injury Criteria

The head injury criteria (HIC) is based on translational accelerations of the head. It was adopted by the U.S. federal government in the Federal Motor Vehicle Safety Standards No. 208. Permissible value of HIC15 is 700. HIC formula is given by [6]:

$$\mathrm{HIC} = \mathrm{Max}\left\{(t_2 - t_1)\left[\frac{\int_{t_1}^{t_2} a(t)dt}{(t_2-t_1)}\right]^{2.5}\right\}, \tag{2}$$

where $a(t)$ is the dummy's head resultant acceleration and $(t_2 - t_1)$ is the time interval. Most commonly used is HIC15 what means that time interval equal 15 ms. With known HIC value it is possible to evaluate the probability of skull fracture [8] :

$$p(\mathrm{HIC}) = \frac{1}{\sqrt{2\pi}}\int_{-\infty}^{\frac{\ln(\mathrm{HIC})-6.96352}{0.84664}} e^{-\frac{t^2}{2}}dt \tag{3}$$

3 FEM Analysis

Four numerical non-linear dynamic analysis crash tests were carried out using LS-DYNA software. The first three describe frontal vehicle accidents with rigid lighting pole at different impact velocity: 30, 50 and 80 km/h. The last one is the full-scale TB11 (100 km/h, 20 degree impact angle) crash test with concrete barrier. The results from both indirect and direct described methods are compared. Finite Element Method model consist of: GeoMetro modified vehicle (including Hybrid III 50th dummy, seatbelts, 60 dm^3 airbag, seat and steering column) and different obstacles as described below.

3.1 Obstacles

3.1.1 Rigid lighting pole

Lighting pole used in frontal collision is 9.7 m high. It is made of shell elements. Rigid material model is used due to improve calculation time. Translation motions and rotations are blocked in lowermost cross section nodes.

3.1.2 Concrete barrier

In the current study, concrete barrier made of 8 meters length segments is used. Barrier consist of 6 segments tied by steel connectors. Obstacle is free-standing on planar ground. Elasto-plastic steel reinforcement is included (beam elements constrained in solid elements). Continuous surface cap material model (CSCM) is used to model concrete behaviour.

3.2 Modified GeoMetro vehicle

Geo Metro (total mass 900 kg) is the smallest vehicle used in standard crash tests [4, 5]. Freely available model was modified by adding simple rigid seat and steering column. In addition, the folded airbag is implemented using control volume formulation. It is made of shell elements which gives 60 dm^3 volume after filling. The three-point seatbelts are also included. Contact parts of the seatbelts is made of shell elements, the rest are made of beam elements. Seatbelt includes pre-tensioner and retractor elements to keep tension after collision. Airbag and seatbelt sensors are triggered by time control, after fulfilling the geometric condition. When sensor is triggered, the seatbelt is going to lock and pretension is induced with 12ms delay. In addition, airbag is filled with 17ms delay. Evaluation of modifications of Geo Metro have been presented in [10].

3.3 Hybrid III 50th dummy

The FEM anthropomorphic male model of a Hybrid III 50th dummy is used to obtain occupant responses in tests. It is made of elastic skeleton covered by rubber parts, viscous foams and viscoelastic skin. The accelerometers are located in the head, chest and pelvis. The family of Hybrid III dummies are being developed at NCAC [9].

4 Results

4.1 Frontal crash tests

The visualization and numerical graphs are showed for frontal collision with rigid lighting pole at 50km/h velocity. Visualization of the initial 0.2 seconds of collision is shown in fig. 1. Fig. 2 shows the dummy's head acceleration and HIC15 calculation range. The time history of ASI is shown in fig. 3. Qualitative character of ASI and head acceleration curves is comparable. Head velocity (V_r), compared to theoretical value (THIV), is presented in fig. 4.

Fig. 1. 50 km/h frontal test visualization.

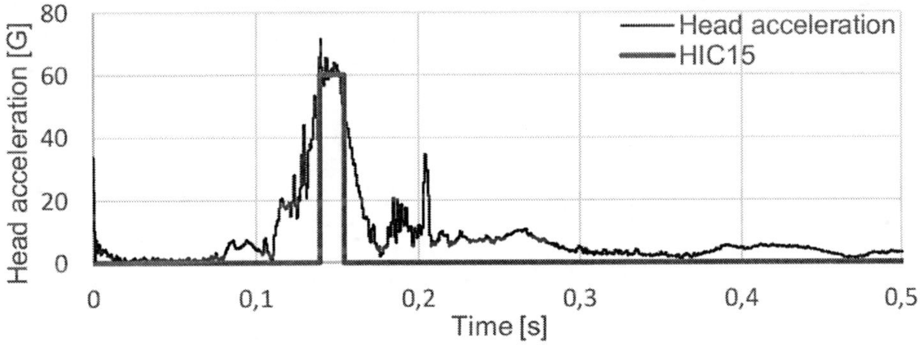

Fig. 2. 50 km/h frontal test, head resultant acceleration and HIC15 calculation range.

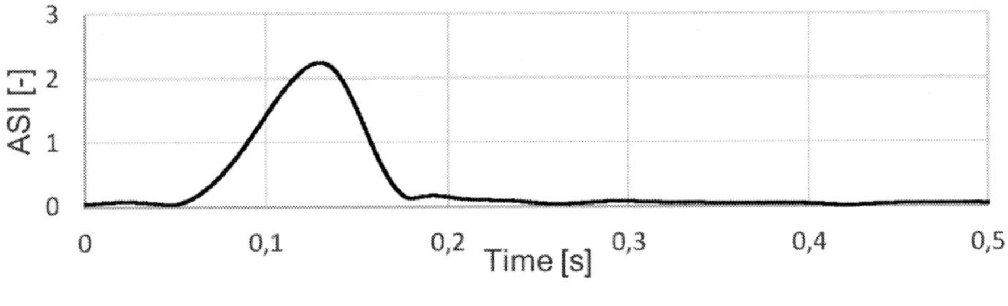

Fig. 3. 50 km/h frontal test, ASI.

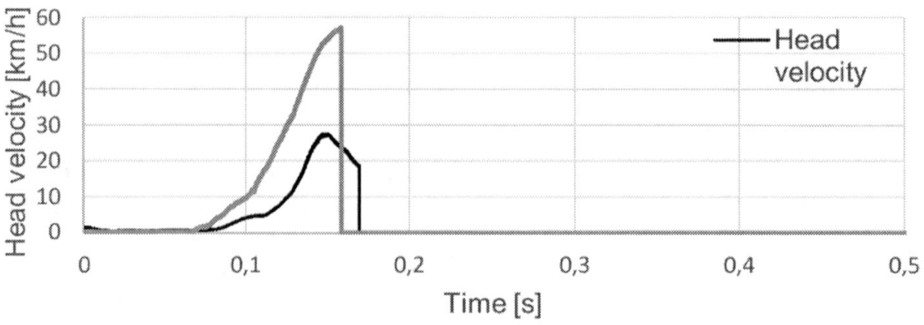

Fig. 4. 50 km/h frontal test, head velocity until first peak Vr and THIV.

Table 5 shows the results of HIC15, head velocity prior to the first peak, Vr, ASI and THIV values for frontal crash test at three different velocities.

Table 5. Results for frontal crash test.

Impact velocity [km/h]	HIC15 [-]	ASI [-]	Vr [km/h]	THIV [-]
30	43.32	1.14	16.8	35.7
50	419.3	2.24	27.4	57.1
80	2454	3.97	67.7	84.1

4.2 TB11 concrete barrier crash test

The initial 0.3 seconds of visualization of crash test simulation with concrete barrier at velocity equal 100 km/h and 20 degree impact angle are shown in fig. 5. Table 6 shows the results of HIC15, head velocity until first peak Vr, ASI and THIV values for TB11 crash test with concrete barrier.

Table 6. Results for TB11 concrete barrier crash test.

Impact velocity [km/h]	HIC15 [-]	ASI [-]	Vr [km/h]	THIV [-]
100	17.1	1.53	21.1	27.0

5. Conclusions

Table 7 shows the summary of the occupant injury estimation results using both methods. It contains probability of skull fracture calculated from HIC15 value, AIS code chosen on the basis of the HIC15 value, injury level based on the AIS code and impact severity level determined on the basis of ASI and THIV values. The D value of the impact severity level means that the results are out of scale.

 https://doi.org/10.1051/matecconf/201928500022

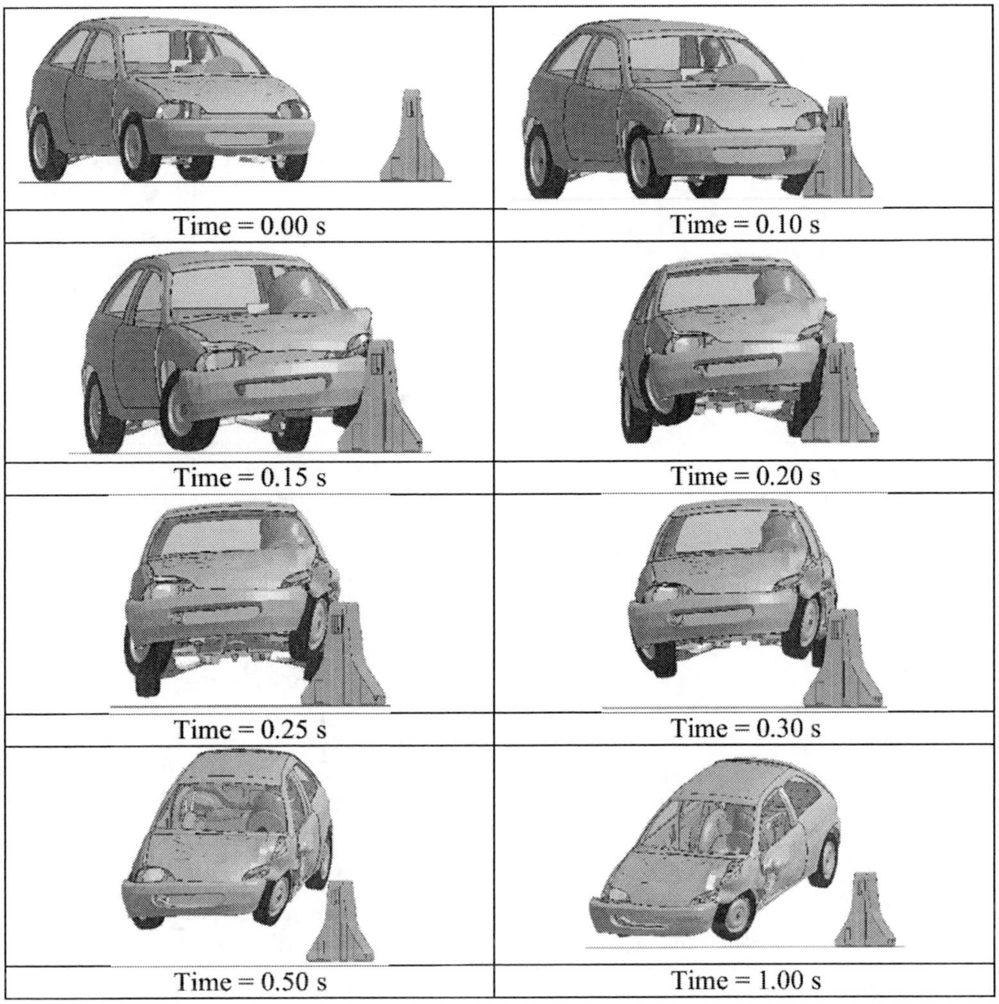

Fig. 5. TB11 crash test visualization.

The results confirm that the indirect method commonly used in the design of barriers has a large safety reserve. In the analysed frontal collision test, directly measured quantities (HIC15, V_r) predicts lower injury risk than those obtained by indirect methods (ASI, THIV). Even for a velocity equal to 50 km/h, the HIC15 corresponds to AIS level 1 (see table 4). Also THIV value is always higher than direct head velocity, which is affected by airbag deployment.

Concrete barrier TB11 test gives acceptable C impact severity index. There is very low value of direct method parameters that means there is no predicted injury. This confirms that the use of the indirect method gives very safe results.

Acknowledgements

This work was supported by the National Centre for Research and Development (NCBiR) and General Director for National Roads and Motorways (GDDKiA) under the research project "Road Safety Equipment" (contract number DZP/RID-I-67/13/NCBR/2016).

Calculations have been carried out at the Academic Computer Centre in Gdańsk, Gdańsk University of Technology, Poland.

Table 7. Summary of the occupant injury estimation.

Method:	**Direct**			**Indirect**
Impact velocity [km/h] / angle [deg]	**p(HIC15) [%]**	**AIS code**	**Injury level**	**Impact severity level**
Frontal collision				
30 / -	0.008	0	No injury	D
50 / -	13.7	1	Minor	D
80 / -	84.0	6	Maximum	D
Concrete barrier				
100 / 20	0.00006	0	No injury	C

References

1. L. Pelenyte-Vysniauskiene, A. Jurkauskas, Transport - 2007, **22**, 269-274 (2007)
2. Sheldon, L.; Stucki, W.; Hollowell, T. Determination of frontal off set test conditions based on crash data. National Highway Traffic Safety Administration, Information Systems and Services. Paper No. 98-S l-0-02. p. 15–35, (1998)
3. Versace, J., A review of the severity index, SAE Technical paper no. 710881; (1971)
4. EN 1317-2 : Road restraint systems – Part 2: Performance classes, impact test acceptance criteria and test methods for safety barriers including vehicle parapets
5. EN 1317-1 : Road restaint systems – Part 1: Terminology and general criteria for test methods
6. C. Andreescu, A. Clenci, Proceedings of the European Automotive Congress EAEC-ESFA, 723-731, (2016)
7. NHTSA, Federal Motor Vehicle Safety Standards : Occupant Crash Protection. 49 C.F.R., Part 571.208, (1998)
8. Hertz E. A note on the head injury criteria (HIC) as a predictor of the risk of skull fracture. In : The 37th annual proceedings of the association for the advancement of automotive medicine, (1993)
9. P. Mohan, C-K. Park, D. Marzougui, C-D. Kan, S. Guha, C. Maurath, et al., LSTC/NCAC dummy model development. In : The 11th international LS-DYNA users conference, Detroit, USA, (2010)
10. Bruski D, Burzyński S, Chróścielewski J, Pachocki Ł, Witkowski W. On the validation of the LS-DYNA Geo Metro numerical model. MATEC Web Conf (2019); 262:10001.
11. EN-12767 Passive safety of support structures for road equipment. Requirements and test methods

Parametric vibrations in mechanical systems on the example of a mining hoist installation

Stanisław Wolny *

AGH University of Science and Technology, Kraków, Poland

Abstract. This study summarises the research efforts undertaken to establish how variable stiffness of shaft steelwork should affect the amplitudes of horizontal displacements of selected components of a conveyance (skip head and skip bottom frame). Most studies investigating the dynamic behaviour of conveyances travelling in the shaft rely on a simplifying assumption whereby the shaft steelwork stiffness is taken to be constant, despite its cyclic variations. This simplification vastly facilitates further dynamic analyses, yet it is necessary to establish which parameters of the shaft steelwork behaviour should be affected, in the context of stiffness in the guiding systems. With regard to parameters of typical hoist installations currently operated in Polish collieries, the study demonstrates that the approach whereby the variable shaft steelwork and its impacts on amplitudes of horizontal displacements of selected components of a conveyance (skip head or bottom frame) should be omitted in the analysis of the hoist dynamics is fully justified and merited in practical applications.

1 Introduction

The dynamic behaviour of a conveyance was analysed under the normal operating conditions (conveyance ride at the fixed velocity V=const) basing on the assumption that horizontal displacements of the lumped masses in the hoist model are attributable to random misalignments and irregularities of the guide strings. Finally, the spectral densities of displacements of the selected mass components were obtained [1].

The formulas derived accordingly and their graphical interpretation [2, 3] should enable the selection of the system's parameters such that displacements of the selected points in the model should not exceed the critical levels specified in relevant mining regulations [4].

The dynamic analysis will yield the variance of interaction forces between a conveyance and shaft steelwork, which is requisite for effective control of the conveyance's profile in the context of its endurance parameters. In order that the results could be reliably used in practical applications, the Author undertook to validate the simplifying assumptions discussed in [1, 2, 3]. The underlying assumption in the dynamic analysis summarized in [1, 2, 3] is the linear dependence of the damping factor (2h) and stiffness (k) of the guiding system (roller guides-shaft steelwork), see Fig 1. The guiding system comprises roller

* Corresponding author: stwolny@agh.edu.pl

guides which, during the conveyance ride, slide on the series- connected guiding rails (beams arranged vertically and supported on buntons (horizontal beams fixed to the shaft steelwork). When the horizontal displacements of a conveyance are insignificant, the elasticity and damping factors are taken to be linear. However, when the guide rails-bunton connections are sufficiently rigid, the guide string can be treated as a beam on elastic supports, its span length l_o equal to the distance between buntons. The tenable assumption is [5] that elasticity of the shaft steelwork (buntons plus the guide rails) should be governed by the formula:

$$k_z = \frac{k_d + k_p}{2} - \frac{k_d - k_p}{2} \cos \omega t \tag{1}$$

where:
k_z- equivalent elasticity factor of the shaft steelwork
k_d- equivalent elasticity factor of shaft steelwork at the point the guide rail is fixed to the bunton;
k_p-equivalent elasticity factor of the shaft steelwork at mid-span between two buntons;
ω - frequency

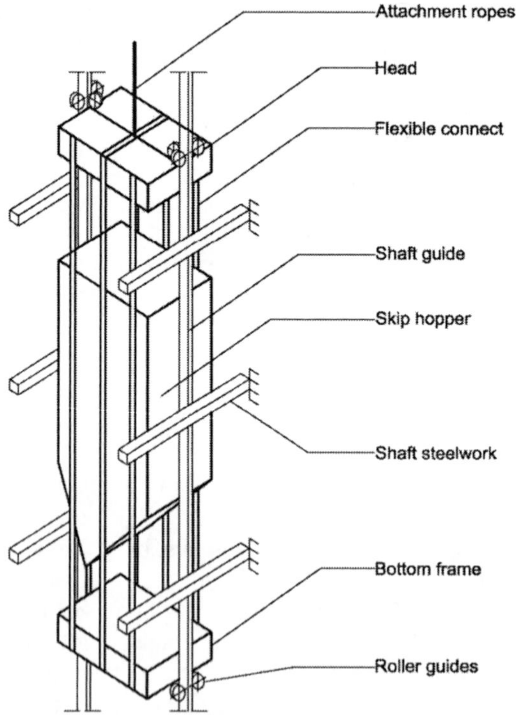

Fig 1. Schematic diagram of the hoist installation.

The assumption whereby the elasticity factor of the shaft steelwork is taken to be constant irrespective of periodic variations (1) simplifies further analysis yet it is necessary to establish to which parameters of the shaft steelwork behaviour it should be applicable in the context of rigidity in the guiding systems.

The impacts of varying shaft steelwork elasticity/stiffness on stability of vibrations in the conveyance-shaft steelwork system were investigated in the work [6]. This study summarises the Author's efforts to determine how variable stiffness of the shaft steelwork should affect the amplitude of horizontal displacements of selected components of the conveyance (for example roller guides).

2 Analysis of dynamic behaviour of the conveyance [1, 2, 3]

In order to establish how variable stiffness of the shaft steelwork should affect the amplitudes of horizontal displacements of conveyance components during its ride in the shaft, the Author recalls the results of dynamic analyses summarised in the works [1, 2, 3].

The conveyance is represented by a triple mass model (Fig. 2) whose dynamic motion equations are based on an assumption that horizontal displacements of the modelled lumped masses are induced by random irregularities and misalignments of guide strings (indicated as $x_1(t)$ and $x_1(t+\tau)$ in Fig. 2).

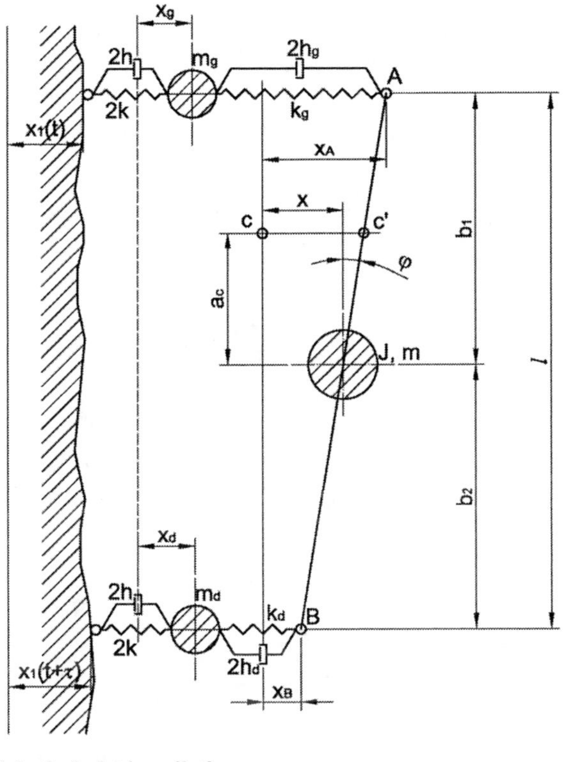

Fig 2. Mechanical model of a hoist installation.

Underlying the equations of motion of the system are the Lagrange equations of the second type. Accordingly, we get [1, 2, 3]:

$$\frac{mb_2^2+J}{l^2}\ddot{x}_A+\frac{mb_1b_2-J}{l^2}\ddot{x}+k_g(x_A-x_g)+2h_g(\dot{x}_A-\dot{x}_g)=0 \quad (2)$$

$$\frac{mb_1^2+J}{l^2}\ddot{x}_B+\frac{mb_1b_2-J}{l^2}\ddot{x}+k_d(x_B-x_d)+2h_d(\dot{x}_B-\dot{x}_d)=0 \quad (3)$$

$$m_g\ddot{x}_g+2k\left(x_g-x_1(t)\right)-k_g(x_A-x_g)-2h_g(\dot{x}_A-\dot{x}_g)+2h\left(\dot{x}_g-\dot{x}_1(t)\right)=0 \quad (4)$$

$$m_d\ddot{x}_d-k_d(x_B-x_d)+2k\left(x_d-x_1(t+\tau)\right)-2h_d(\dot{x}_B-\dot{x}_d)+2h\left(\dot{x}_d-\dot{x}_1(t+\tau)\right)=0 \quad (5)$$

where:

$x_A=x+b_1\varphi$, $x_B=x+b_2\varphi$,
φ- angle of skip hopper rotation around its centre of gravity (c.o.g)
x- horizontal displacement of a hopper mass
m- mass of a skip hopper plus the payload
J- inertia moment of a skip hopper with payload
m_g- mass of the skip head
m_d- mass of the bottom frame
2h, k- (linear) damping and elasticity factors of in the guiding system (roller guides- shaft steelwork)
$2h_g$, k_g- damping and elasticity factors of (lateral) flexible connector between the head and hopper
$2h_d$, k_d- damping and elasticity factors of (lateral) flexible connectors between the bottom frame and skip hopper
x_g, x_d,- horizontal displacements of the face of the skip head and bottom frame, respectively
$x_1(t)$, $x_1(t+\tau)$- function defining the imperfections of cage guides' front surfaces
τ- time of conveyance ride along the path equal to the distance between the front shoes (on the head and the bottom frame)
b_1,b_2- distance between the hopper c.o.g and front shoes on the top and at the bottom, respectively
l- distance between top and bottom shoes

Recalling the work of Kawulok [2], it is assumed that irregularities of the guide string $x_1(t)$ and $x_1(t+\tau)$ can be treated as a random stochastic process, its spectral density expressed as:

$$S_x=\frac{2D_{x_1}\alpha}{\alpha^2+\omega^2}\ [\mathrm{m^2s}] \tag{6}$$

where: D_{x_1}, α- respective parameters.

Basing on the results of dynamic analysis, the variance of amplitude of the conveyance-shaft steelwork interaction forces can be derived accordingly [5, 6, 7]. In order that the results could be utilised in practical applications, the Author resolved to validate the simplifying assumptions discussed in [5, 6, 7]. In this study the Author focused on establishing the impact of variable shaft steelwork stiffness on amplitudes of horizontal displacements of selected components of the conveyance.

3 Induced parametric vibrations

The system of equations (2)-(5) is recalled to evaluate the impacts of variable stiffness of the shaft steelwork (Eq. 1) on amplitude of horizontal displacements of the conveyance during its ride at fixed velocity V_0=const. The term expressing the constant elasticity of the guiding system k in the system of equations (2)-(5) is replaced by the parameter k_c (expressing the sum total of the shaft steelwork elasticity k_z and that of the roller guide k_k).

In this study the evaluation procedure is limited to a specific case whereby the condition $J=mb_1b_2$ is satisfied and hence the system of equations (2)-(5) could be uncoupled yielding two independent systems of equations (2)-(3) and (4)-(5).

Accordingly, the system of equations (2)-(3) is rewritten as:

$$\begin{gathered} m_A\ddot{x}_A+k_g(x_A-x_g)+2h_g(\dot{x}_A-\dot{x}_g)=0 \\ m_g\ddot{x}_g+2k_c[x_g-x_1(t)]-k_g(x_A-x_g)-2h_g(\dot{x}_A-\dot{x}_g)+2h[\dot{x}_g-\dot{x}_1(t)]=0 \end{gathered} \tag{7}$$

where

$m_A=\frac{J+mb_2^2}{l^2}$,

k_c- total elasticity factor of the whole system comprising the roller guides and shaft steelwork.

The total elasticity factor of the whole system comprising the roller guides and shaft steelwork (for the guiding elements connected in series) becomes:

$$k_c=\frac{E\cdot C(1-\frac{D}{C}\cos\omega t)}{(E+C)(1-\frac{D}{E+C}\cos\omega t)}, \tag{8}$$

where

$\frac{k_d+k_p}{2}=C$; $\frac{k_d-k_p}{2}=D$; $k_k=E$ - elasticity factor of the roller guide.

After respective substitutions we get:

$$k_p=\frac{E\cdot C}{E+C};\ \mathcal{E}_1=\frac{D}{E\cdot C};\ \mathcal{E}_2=\frac{D}{E+C},$$

finally, (8) becomes:

$$k_c=k_p\frac{1-\varepsilon_1\cos\omega t}{1-\varepsilon_2\cos\omega t}, \tag{9}$$

The values of $\mathcal{E}_1$, $\mathcal{E}_2$ are much less than one, accordingly, the formula (9) can be rewritten as:

$$k_p\frac{1-\mathcal{E}_1\cos\omega t}{1-\mathcal{E}_2\cos\omega t}\approx k_p(1+\varepsilon\cos\omega t) \tag{10}$$

where:

$\varepsilon=\mathcal{E}_2-\mathcal{E}_1$ or $\varepsilon=\frac{D}{C}-\frac{D}{E+C}=\frac{ED}{C(E+C)}$.

Finally, (9a) becomes:

$$k_c=k_p(1+\varepsilon\cos 2\pi\frac{V_0}{l_0}t) \tag{11}$$

where:

V_0 [m/s]-steady -state velocity of the conveyance travel;

l_0 [m]- distance between bunton;

The underlying assumption is that vibrations of the modelled system governed by the equation (7) have been induced by kinematic displacement inputs

$x_1(t)=B\cdot f_{(t)}$,

where:

$f_{(t)}=f_0+f_2\cos(2\pi\frac{V_0}{l_0}t)$

Substituting (11) into the system of equations (7) yields:

$$\begin{gathered}m_A\ddot{x}_A+k_g(x_A-x_g)+2h_g(\dot{x}_A-\dot{x}_g)=0\\ m_g\ddot{x}_g+2k_p\left(1+\varepsilon\cos 2\pi\frac{V_0}{l_0}t\right)x_g-2k_p\left(1+\varepsilon\cos 2\pi\frac{V_0}{l_0}t\right)-2h_p\left(1+\varepsilon\cos 2\pi\frac{V_0}{l_0}t\right)+x_1(t)+\\ -k_g(x_A-x_g)-2h_g(\dot{x}_A-\dot{x}_g)+2h\dot{x}_g+2h\frac{dx_1}{dt}=0\end{gathered} \tag{12}$$

In terms of dimensionless parameters, $\tau = p_{10}t$ in the system of equations (12), where p_{10} is the natural vibration frequency of the mass m_A (for m_g=0). Substituting:

$$\begin{gathered}X_A(t)=u(\tau),\\ X_g(t)=\upsilon(\tau),\\ X_1(t)=\omega(\tau)=B[f_0+f_2\cos(2\pi\frac{V_0}{l_0}\cdot\frac{\tau}{p_{10}})],\end{gathered} \tag{13}$$

rearranging and dividing the first equation in the system of equations (12) by $m_A p_{10}^2$ and the second equation by $m_g p_{10}^2$, we obtain:

$$\ddot{u}+\frac{2\mathrm{h_g}}{\mathrm{m_A p_{10}}}\dot{u}+\frac{\mathrm{k_g}}{\mathrm{m_A p_{10}}^2}u-\frac{2\mathrm{h_g}}{\mathrm{m_A p_{10}}}\dot{v}-\frac{\mathrm{k_g}}{\mathrm{m_A p_{10}}^2}v=0 \tag{14}$$

$$\ddot{\upsilon}+\frac{2(h+h_g)}{\mathrm{m_g}\cdot\mathrm{p_{10}}}\dot{\mathrm{v}}+\left[2\mathrm{k_p}\left(1+\mathcal{E}\cos 2\pi\frac{V_0}{l_0}\frac{\tau}{\mathrm{p_{10}}}\right)+\mathrm{k_g}\right]\cdot\frac{1}{\mathrm{m_g p_{10}}^2}\upsilon-\frac{2h_g}{\mathrm{m_g p_{10}}}\dot{u}-\frac{\mathrm{k_g}}{\mathrm{m_g p_{10}}}u+\frac{2\mathrm{h_g}}{\mathrm{m_A p_{10}}}\cdot$$
$$\cdot B\cdot f_2\cdot 2\pi\frac{V_0}{l_0}\frac{1}{\mathrm{p_{10}}}\sin\left(2\pi\frac{V_0}{l_0}\cdot\frac{\tau}{\mathrm{p_{10}}}\right)-2\mathrm{k_p}\left(1+\mathcal{E}\cos 2\pi\frac{V_0}{l_0}t\right)B[f_0+f_2\cos(2\pi\frac{V_0}{l_0}\cdot\frac{\tau}{\mathrm{p_{10}}})]\cdot\frac{1}{\mathrm{m_g p_{10}}^2}=0$$

After respective substitutions

$$n_1=\frac{\mathrm{m_A}}{\mathrm{m_g}};\ n_2=\frac{h}{\mathrm{m_g}\cdot\mathrm{p_{10}}};\ n_3=\frac{\mathrm{k_p}}{\mathrm{k_g}};\ n_4=2\pi\cdot\frac{\upsilon_0}{\mathrm{l_0}}\cdot\frac{1}{\mathrm{p_{10}}};\ \mathrm{p_{10}}=\sqrt{\frac{\mathrm{k_g}}{\mathrm{m_A}}},$$
$$\frac{h_g}{h}\ll 0,\ h_g+h\cong h\ \text{ and }\ h_g=0$$

We get

$$\ddot{\mathrm{u}}+\mathrm{u}-\upsilon=0$$
$$\ddot{\upsilon}+2n_2v+n_1[1+2n_3(1+\mathcal{E}\cos n_4\tau)]v-n_1u=2n_1n_3B(1+\mathcal{E}\cos n_4\tau)(f_0+f_2\cos n_4\tau)+$$
$$-2n_2n_4Bf_2\sin n_4\tau \tag{15}$$

To obtain the maximal amplitudes of displacement, damping in the system is neglected (n_2=0) and hence the system of equations (15) becomes:

$$\ddot{\mathrm{u}}+\mathrm{u}-\upsilon=0$$
$$\ddot{\mathrm{v}}+\mathrm{n_1}[1+2\mathrm{n_3}(1+\mathcal{E}\cos\mathrm{n_4}\tau)]\upsilon-\mathrm{n_1 u}=\mathrm{f_{01}}+\mathrm{f_{02}}\cos\mathrm{n_4}\tau+\mathrm{f_{03}}\cos 2\mathrm{n_4}\tau \tag{16}$$

where:

$$f_{01}=n_1n_3B(2f_0+\mathcal{E}f_2),\ f_{02}=2n_1n_3B(f_0+f_2),\ f_{03}=n_1n_3Bf_2\cdot\mathcal{E}.$$

Solution to the approximate system of equations is given as [7]:

$$u=A_0+A_1\sin n_4\tau+B_1\cos n_4\tau \tag{17}$$
$$\upsilon=B_0+A_2\sin n_4\tau+B_2\cos n_4\tau \tag{18}$$

Respective values of these parameters are:

$$A_0=B_0=\frac{f_{01}}{2n_1n_3}-\frac{\mathcal{E}}{2}\frac{f_{02}-\mathcal{E}f_{01}}{n_1+n_1n_3\left(2-\mathcal{E}^2\right)-n_4^2}\cdot\frac{\mathrm{m}}{\mathrm{p}} \tag{19}$$
$$A_1=n_4^2\frac{n_1+2n_1n_3-n_4^2}{2n_1n_3-n_4^2}\cdot\frac{f_{02}-\mathcal{E}f_{01}}{n_1+n_1n_3\left(2-\mathcal{E}^2\right)-n_4^2}\cdot\frac{1}{\mathrm{p}} \tag{20}$$
$$A_2=\frac{n_1n_4^2}{2n_1n_3-n_4^2}\cdot\frac{f_{02}-\mathcal{E}f_{01}}{n_1+n_1n_3\left(2-\mathcal{E}^2\right)-n_4^2}\cdot\frac{1}{\mathrm{p}} \tag{21}$$
$$B_1=\frac{f_{02}-\mathcal{E}f_{01}}{n_1+n_1n_3\left(2-\mathcal{E}^2\right)-n_4^2}\cdot\frac{1}{\mathrm{p}} \tag{22}$$
$$B_2=\frac{f_{02}-\mathcal{E}f_{01}}{n_1+n_1n_3\left(2-\mathcal{E}^2\right)-n_4^2}\cdot\frac{\mathrm{m}}{\mathrm{p}} \tag{23}$$

where:

$$\mathrm{p}=\mathrm{m}-\frac{n_1}{n_1+n_1n_3\left(2-\mathcal{E}^2\right)-n_4^2},\ \ \mathrm{m}=1-n_4^2\frac{n_1+2n_1n_3-n_4^2}{2n_1n_3-n_4^2}.$$

Accordingly, the solution becomes:

$$\mathrm{u}=A_0+A_1\sin n_4\tau+B_1\cos n_4\tau=$$
$$=\frac{f_{01}}{2n_1n_3}-\frac{\mathcal{E}}{2}\frac{f_{02}-\mathcal{E}f_{01}}{n_1+n_1n_3\left(2-\mathcal{E}^2\right)-n_4^2}\cdot\frac{\mathrm{m}}{\mathrm{p}}+\frac{f_{02}-\mathcal{E}f_{01}}{n_1+n_1n_3\left(2-\mathcal{E}^2\right)-n_4^2}\cdot\frac{\mathrm{q}}{\mathrm{p}}\sin[n_4\tau+\mathrm{arctg}\varphi] \tag{24}$$

and

$$\upsilon=B_0+A_2\sin n_4\tau+B_2\cos n_4\tau=$$

$$=\frac{f_{01}}{2n_1n_3}-\frac{\varepsilon}{2}\frac{f_{02}-\varepsilon f_{01}}{n_1+n_1n_3\left(2-\varepsilon^2\right)-n_4^2}\cdot\frac{\mathrm{m}}{\mathrm{p}}+\frac{f_{02}-\varepsilon f_{01}}{n_1+n_1n_3\left(2-\varepsilon^2\right)-n_4^2}\cdot\frac{\mathrm{r}}{\mathrm{p}}\sin[n_4\tau+\mathrm{arctg}\psi] \quad (25)$$

where:

$$\mathrm{q}=\sqrt{\left[n_4^2\frac{n_1+2n_1n_3-n_4^2}{2n_1n_3-n_4^2}\right]^2+1};\quad \mathrm{r}=\sqrt{\left[\frac{n_1n_4^2}{2n_1n_3-n_4^2}\right]^2+\mathrm{m}^2} \quad (26)$$

$$\varphi=\mathrm{arctg}\left[\frac{1}{n_4^2}\frac{2n_1n_3-n_4^2}{n_1+2n_1n_3-n_4^2}\right];\quad \psi=\mathrm{arctg}\left[\mathrm{m}\frac{2n_1n_3-n_4^2}{n_1n_4^2}\right]. \quad (27)$$

The impacts of variable stiffness of the shaft steelwork on amplitude of horizontal displacements of the skip head under parametric vibration excitations are illustrated in Fig 3,4.

Fig. 3 plot the amplitudes of skip head displacement (υ) as a function ε=0.01; 0.1; 0.5 expressing the periodically varying shaft steelwork stiffness, given the remaining dimensionless parameters n_1,n_3,n_4 of the investigated system (damping n_2 has been neglected to obtain the maximal displacement values).
The range of numerical values of parameters n_1,n_3 and n_4 was chosen such that they should be relevant and applicable to typical hoisting installations currently designed and operated in Poland [8].

In accordance with relevant mining regulations [4, 5], the respective values of parameters of the applied excitations considered in the analysis were taken as follows: B=5 mm, f_0=1, f_2= ε,

4 Conclusions

As mentioned in earlier sections, displacements in the conveyance guiding system are induced by the conveyance-shaft steelwork interaction forces, which are of key importance in the context of fatigue endurance of the load-bearing components in the system.
Within the framework of endurance analysis, at the stage of dimensioning the magnitudes and configuration of all loads acting upon the designed structure should be precisely determined. That means that the physical model underlying the stress analysis ought to capture all relevant aspects, including the impacts of varying shaft steelwork stiffness on amplitudes of horizontal displacements of the skip components (skip head and bottom frame), which were omitted in the dynamic analyses summarised in the works [2, 3]. Plots in Fig. 3 and 4 indicate that the omission of variable shaft steelwork stiffness and its impacts on skip head displacements is revealed for $\varepsilon\geq0.1$ whilst in typical hoisting installations currently operated in Poland the values of this parameter are less than 0.01 ($\varepsilon\leq0.01$) [8, 9].

Thus, that the approach whereby the varying shaft steelwork stiffness and its impacts on amplitudes of horizontal displacements of selected skip components (skip head, bottom frame) should be omitted in the analysis of hoist dynamics appears to be fully justified and merited in practical applications.

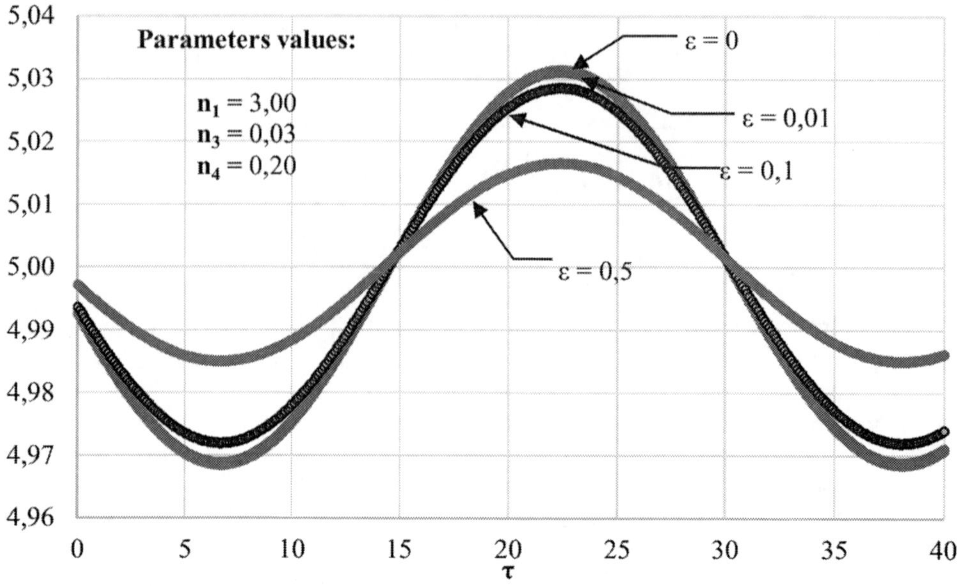

Fig 3. Selected sections of the plot of the skip head displacement $\upsilon(\tau)$ in the function ε, for $n1$=3.0; $n3$=0.03; $n4$=0.2.

References

1. S. Wolny, *Loads experienced by load-bearing components of mine hoist installations due to random irregularities and misalignments of the guide string*, Journal of Machine Construction and Maintenance, vol. **110**, iss. 3,p.79-86 (2018)
2. S. Wolny, *Interactions in Mechanical Systems Due to Random Inputs on the example of a Mine Hoist*, International Education and Research Journal, vol. **1**. iss.5,p.72-76 (2015)
3. S. Wolny, *Displacements in Mechanical System Due to Random Inputs in a Mine Hoist Installation*, Engineering Transactions. Engng. Trans. **65**,3,p.513-522 (2017)
4. Mining Regulation: The Council of Ministers regulation of 30 April 2004 in the admission of using in mining plants (Dz.U. No. 99,item 1003 of 2005 No. 80, item 695 and of 2007 No. 249, item 1853, p.1.2 skip hopper).
5. S. Kawulok, *The influence of the shaft steelwork on mechanics of guiding skip hopper* , GIG works, Katowice (1989)
6. S. Wolny, A. Drzewosz, *Stability in mechanical system on the example of mine hoist installation* (Article submitted for publication in Archives of Mining Sciences) (2018)
7. R. Gutowski, W. A. Swietlicki, *Dynamics and vibrations of mechanical systems*, PWN, Warsaw (1986)
8. S. Wolny et al., *Developing critiria for assessing the supporting structure skip hopper of hoist installation in the terms of extending the period of safe operation*, Scientific an reasearch work, Department of Strength of Materials and Structures, AGH, Cracow (2003)
9. S. Wolny et al., *Development of multi-option modernization of mining cage lift R-VII*, Scientific and research work, Department of Strength and Fatigue of Materials and Structures, AGH, Cracow (2011)

MATEC Web of Conferences **285**, 00024 (2019) https://doi.org/10.1051/matecconf/201928500024
DYNKON 2019

The effects of irregularities and misalignments of the guide strings on stress levels in selected load-bearing elements of a conveyance

Stanisław Wolny[1,*], and *Agata* Drzewosz[1,*]

[1]AGH university of Science and Technology, Kraków, Poland

Abstract. In order that the ultimate state methods can be applied in dimensioning of the load-bearing elements in a conveyance it is required that their design loads should be determined during their normal duty cycle and under the emergency braking conditions.

This study aims to verify the predicted loading conditions obtained by theoretical analyses through experiments on a real object. Prior to the actual experimental procedure, limited to measurements of stress acting in the load-bearing cables in the conveyance, the system parameters have been identified. Special attention was given to parameters of roller guides and random irregularities of the guiding strings, which give rise to horizontal displacements of the conveyance and generate the conveyance- shaft steelwork interaction forces.

1 Introduction

With the assumption made that horizontal displacements of lumped masses are induced by random irregularities and misalignments of the guide strings[1, 2], the analysis of the skip's dynamic behaviour in normal operating conditions (at fixed hoisting velocity V_0=const) would yield the spectral densities of interaction forces between modelled masses as well as variances of the force amplitudes.

For the adopted calculation model of the skip (a 2D framework diagram), the magnitudes of forces were established that act upon the cross-sections of the load-bearing cables. In the work [3] the bending moments were determined. Thus obtained values of forces acting upon the cable cross-sections allowed the maximal stress levels acting upon a single cable to be determined [3].

The adequacy of the theoretical analyses of system dynamics and of fatigue endurance aspects has been verified through experiments on a real object [4, 5], involving stress and strain measurements in cables. Measurement results are compiled in [6].

The analysis of measurement results reveals a major discrepancy between predicted data and experimental results. The study is limited to comparing and contrasting experimental results with predicated data including:

- conveyance-shaft steelwork interaction forces

* Corresponding author: stwolny@agh.edu.pl, drzewosz@agh.edu.pl

- stress levels in load-bearing cables in the conveyance.

Underlying the theoretical analyses are the following assumptions made:

- physical parameters of the guiding elements (roller guides) are based on the manufacturer's specifications;
- horizontal displacements of modelled masses are induced by irregularities and misalignments of the guide strings.

In order to find the cause of such major discrepancy of results, the Authors first identified the real parameters of roller guides and examined the actual irregularities of the guiding strings.

2 Roller guides are the components of the investigated hoist installation

The schematic diagram of the hoist installation considered in the experimental procedure is shown in Fig 1.

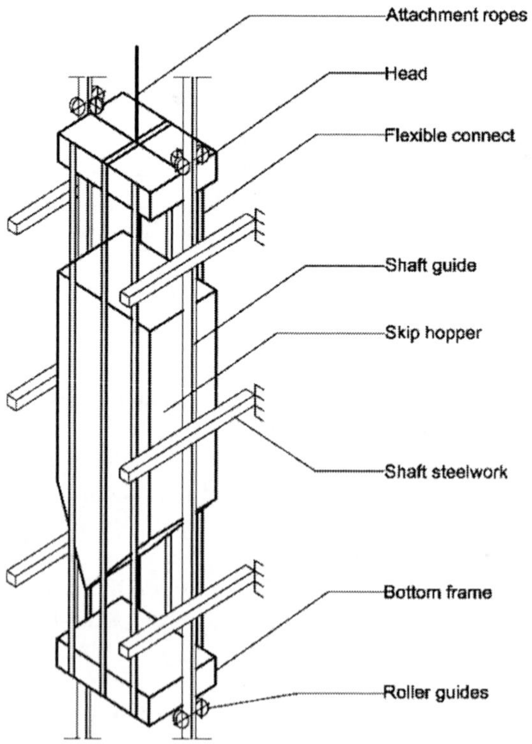

Fig 1. Schematic diagram of the hoist installation [7].

Fig 2 shows the structural design of the guide element mounted on the conveyance [7].

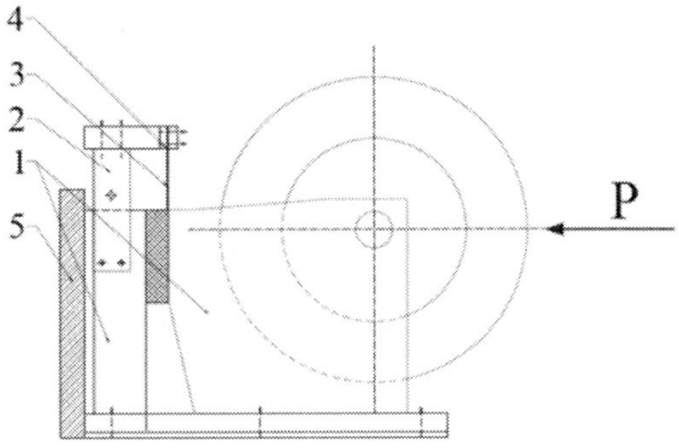

Fig 2. Diagram of the front and lateral guides (1- roller guide, 2- body of the measuring instrument,3- elastic bar, 4- strain gauge, 5-base)[7].

Elasticity factor and viscous damping coefficient of the roller guides were obtained in laboratory tests performed as a part of the research study [7]. These parameters are:

$$k=7{,}5\cdot10^5\left[\frac{N}{m}\right],\ \ 2h=800\left[\frac{N\cdot s}{m}\right].$$

Elasticity of the skip structure in the plane determined by the guides' front surface is derived from the formula [4]:

$$k_g=\frac{96EI_1}{\left(\frac{l_1}{2}\right)^3}, \tag{1}$$

$I_1=\frac{bh^3}{12}$ - inertia moment of the cable with respect to the main, central axis of inertia, perpendicular to the front plane of the skip;

- b, h [m] - dimensions of the cable cross- section;
- l_1 - distance between the front guides on top and at the bottom frame.

Assuming the geometric dimensions of the conveyance to be [4]: b=0,03 m, h=0,25 m, l_1=16 m, we obtain:

$$k_g=\frac{96\cdot2{,}1\cdot10^{11}\cdot0{,}03\cdot0{,}25^3}{12\cdot8^3}=1{,}57\cdot10^6\left[\frac{N}{m}\right].$$

3 Irregularities of the guiding strings

Irregularities and misalignments of the guiding strings in the hoistway were investigated in the shaft in which the experiment was carried out, involving the measurement of conveyance-shaft steelwork interaction forces and of stress levels in load-bearing elements of a conveyance [4, 6]. Guiding strings, shown schematically in Fig 1, exhibit irregularities and certain misalignment, which were the subject of analysis. These irregularities of two guiding strings in the hoistway are designated as fx_1 and fx_4 (for one guide string) and fx_2 and fx_3 (for the other guide string). Measured misalignments of the guide string are plotted in Fig 3 and 4. Fig 3 plots measured misalignment fx_2 and fx_3 (deflection from the vertical) for one guiding string, misalignments of the second string are shown in Fig 4.

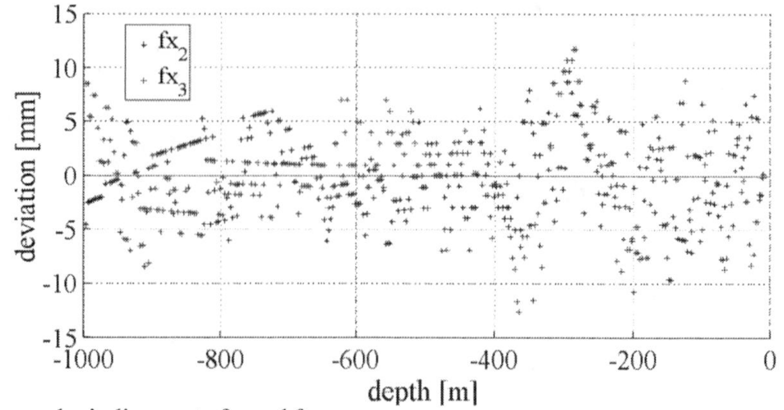

Fig. 3. Measured misalignments fx_2 and fx_3.

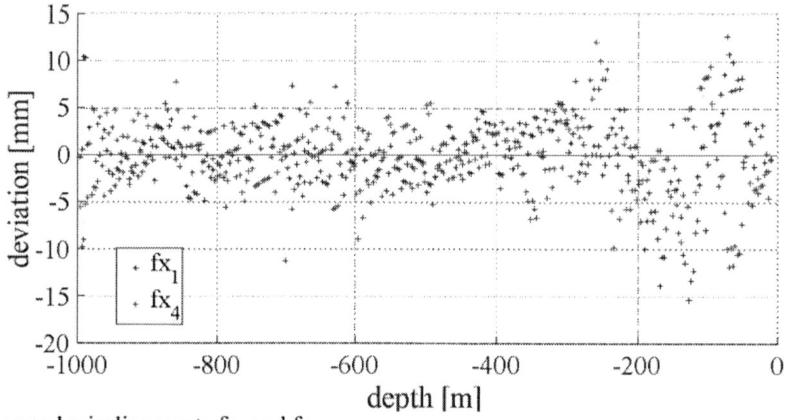

Fig. 4. Measured misalignments fx_1 and fx_4.

Measurement data were used to determine the function that correlates the irregularities of the guiding strings for varied hoisting velocities, given by the formula [8]:

$$R(\tau)=\int_{-\infty}^{\infty} f(t)\bar{f}(t-\tau)dt \qquad (2)$$

where: $f(t)$ – the original signal, $\bar{f}(t-\tau)$ – a correlated signal delayed by τ, $R(\tau)$ – correlation function

Calculating the correlation functions of irregularities and misalignments of the guide strings, we proceed to obtain the spectral density functions of those irregularities (see Fig 5, 6, 7).

Spectral density functions capturing guide string irregularities are plotted separately for each of the guide rails, for three values of hoisting velocity. The maximal values range from $1{,}5mms^2 \div 4mms^2$. A thorough analysis of these plots reveals a certain dependence associated with the actual graph shape. By applying the least square method approximation, the graphs can be substituted by the function [9]:

$$S_x=\frac{2D_x\alpha}{\alpha^2+\omega^2} \qquad (3)$$

Parameters D_x and α were adjusted using the MATLAB software and ω is used to denote frequency given in $\left[\frac{1}{s}\right]$. The analysis was performed for three velocity values

 https://doi.org/10.1051/matecconf/201928500024

$\left(V_0=10\frac{\text{m}}{\text{s}}, 12\frac{\text{m}}{\text{s}}, 16\frac{\text{m}}{\text{s}}\right)$ and three functions obtained accordingly were approximated yielding the spectral density of irregularities of guiding strings (see Fig 5, 6, 7).

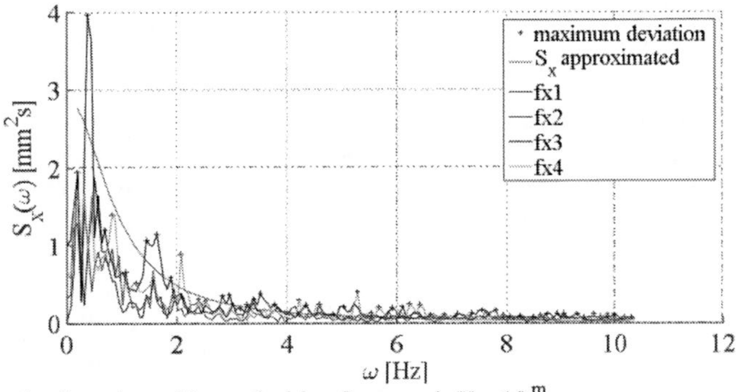

Fig 5. Spectral density function of irregularities for speed $V_0=10\frac{\text{m}}{\text{s}}$.

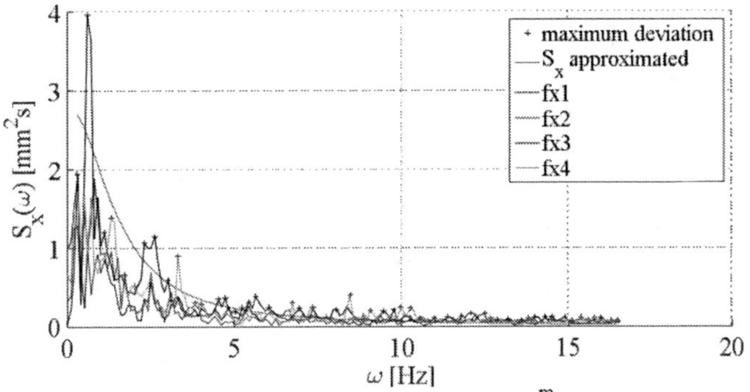

Fig 6. Spectral density function of irregularities for speed $V_0=12\frac{\text{m}}{\text{s}}$.

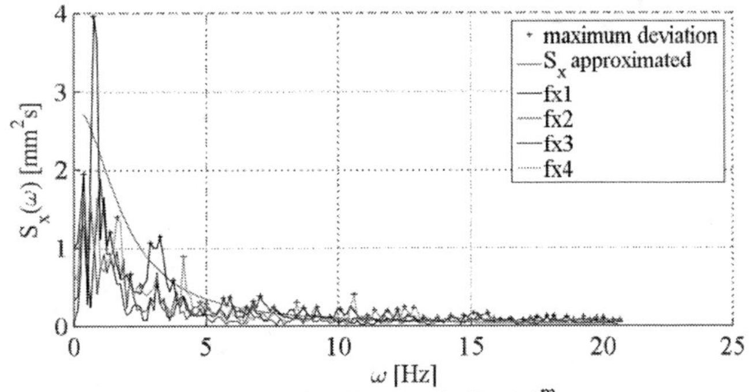

Fig 7. Spectral density function of irregularities for speed $V_0=16\frac{\text{m}}{\text{s}}$.

Table 1. Parameters values.

Velocity V_0 $\left[\frac{\text{m}}{\text{s}}\right]$	Parameters		Formula $S_x=\frac{2D_x\alpha}{\alpha^2+\omega^2}$
	α $\left[\frac{1}{\text{s}}\right]$	D_x [mm²]	
10	0,94	1,32	$S_x(\omega)=\frac{2D_x\alpha}{\alpha^2+\omega^2}=\frac{2\cdot1{,}32\cdot0{,}94}{0{,}94^2+\omega^2}$
12	1,50	2,11	$S_x(\omega)=\frac{2D_x\alpha}{\alpha^2+\omega^2}=\frac{2\cdot2{,}11\cdot1{,}5}{1{,}5^2+\omega^2}$
16	1,87	2,64	$S_x(\omega)=\frac{2D_x\alpha}{\alpha^2+\omega^2}=\frac{2\cdot2{,}64\cdot1{,}87}{1{,}87^2+\omega^2}$

For the steady hoisting velocity like $V_0=10\frac{\text{m}}{\text{s}}$, $V_0=12\frac{\text{m}}{\text{s}}$, $V_0=16\frac{\text{m}}{\text{s}}$, the spectral density function of guide string irregularities and misalignments are written as examples in Table 1.

Plots of spectral density of guide string irregularities for the considered hoisting velocities ($V_0=10\frac{\text{m}}{\text{s}}$, $12\frac{\text{m}}{\text{s}}$, $16\frac{\text{m}}{\text{s}}$) are compiled in respective figures. Parameters D_x and α were adjusted for each hoisting velocity such that the mean square error involved in the approximation procedure should be as small as possible. As the value of spectral density function $S_x(\omega)$ tends to increase with increasing velocity, the function used in further analyses will be that expressed as:

$$S_x(\omega)=\frac{9{,}87}{1{,}87^2+\omega^2}$$

like in the case of $V_0=16\frac{\text{m}}{\text{s}}$.

4 Stress levels in load-bearing cables in a conveyance

Spectral density of stresses acting in load-bearing cables of a conveyance due to irregularities and misalignments of the guide string can be derived from the formula [3, 8]:

$$D_\sigma=\frac{3}{2}D_x\cdot n\cdot k_g^2\cdot\frac{l_1}{bh^2}\cdot I_{50} \tag{4}$$

where:
D_x - parameter characterising irregularities of the guide string;
k_g - elasticity of guide string in the front plane of the skip
l_1- midpoint of the skip height (distance between the front shoes on the skip head and those on the bottom frame)
b, h - cross-section areas of the bearing cables
and

$$I_{50}=\frac{M_{50}}{a_0\Delta_{50}} \tag{5}$$

$$M_{50}=-a_0\{a_5[b_1(a_3a_4-a_2a_5)+b_2(a_0a_5-a_1a_4)+b_3(a_1a_2-a_0a_3)]+b_4[a_1(a_1a_4-a_2a_3)+a_0(a_3^2-a_1a_5)]\}$$

$$\Delta_{50}=-a_5\{a_0^2a_5^2-2a_0a_1a_4a_5-a_0a_2a_3a_5+a_0a_3^2a_4+a_1^2a_4^2+a_1a_2^2a_5-a_1a_2a_3a_4\}$$

$a_0=1$; $a_1=2n_2+n$; $a_2=1+n_1(1+2n_3)+2n_2n$; $a_3=2n_2+n+n_1n(1+2n_3)$; $a_4=2(n_1n_3+n_2n)$; $a_5=2n_1n_3n$; $b_0=0$; $b_1=-n_2^2$; $b_2=4n_1^2n_3^2$; $b_3=0$; $b_4=0$; $n_1=\frac{m_A}{m_g}$; $n_2=\frac{h}{m_g}\sqrt{\frac{m_A}{k_g}}$; $n_3=\frac{k}{k_g}$; $n=\alpha\sqrt{\frac{m_A}{k_g}}$

Since $\frac{h_g}{h}\ll 1$, it is assumed that $h+h_g=h$ and $h_g\cong 0$. The remaining designations are identical to those adopted in [3].

The product $n\cdot D_x$ present in Eq(4) determines the parameters of the applied random excitations whilst the expression (5) is governed by parameters characterising the investigated object.

Variability ranges of parameters n_1, n_2, n_3 and n are adjusted basing on the calculation procedure highlighted in chapters 2 and 3. Parameters used in the procedure were those of the conveyance being the object of experimental testing [4, 7]:

$$m_A=1{,}6\cdot 10^4\ [\text{kg}]\ ;\ m_g=4\cdot 10^3\ [\text{kg}]\ ;$$

Accordingly, we get : $n_1=\frac{m_A}{m_g}=\frac{1{,}6\cdot 10^4}{4\cdot 10^3}=4$; $n_2=\frac{h}{m_g}\sqrt{\frac{m_g}{k_g}}=\frac{4000}{4\cdot 10^3}\sqrt{\frac{1{,}6\cdot 10^4}{1{,}54\cdot 10^6}}\cong 0{,}10$;

$$n_3=\frac{k}{k_g}=\frac{7{,}5\cdot 10^5}{1{,}54\cdot 10^6}\cong 0{,}49\ ;\ n=\alpha\sqrt{\frac{m_A}{k_g}}=1{,}87\sqrt{\frac{1{,}6\cdot 10^4}{1{,}54\cdot 10^6}}\cong 0{,}19$$

Variability of I_{50} depending on $n_1=\frac{m_A}{m_g}$ for n_2=0,10, n_3=0,49 oraz n=0,19 is plotted in Fig. 8.

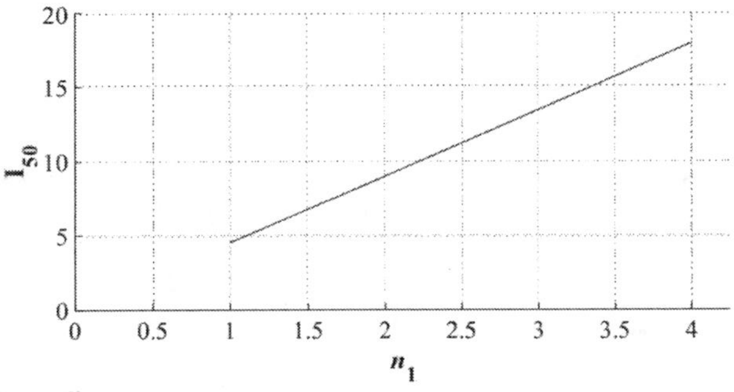

Fig. 8. Diagram of I_{50} depending on n_1.

The value of I_{50} read out from Fig. 8 and obtained for the skip parameters listed above is found to be I_{50}=18.

In accordance with (4), spectral density of stresses in load-bearing cables due to irregularities and misalignment of the guide string becomes:

$$D_\sigma=\frac{3}{2}\cdot 2{,}64\cdot 10^{-6}\cdot 0{,}19\cdot 1{,}57^2\cdot 10^{12}\cdot\frac{16}{0{,}03\cdot 0{,}25^2}\cdot 18=0{,}274\ [\mathrm{MPa}^2]$$

Standard deviation of maximal stresses acting in load-bearing cables in the case analysed in this study is derived from the formula:

$$\sigma(D_\sigma)=\sqrt{D_\sigma}=\sqrt{0{,}274}\cong 0{,}52\ [\mathrm{MPa}]$$

5 Conclusions

Predicted values of standard deviations of maximal stresses acting in load-bearing cables in a skip are decidedly lower than those measured experimentally [4, 6]. Such significant discrepancy (30-fold) can be attributable to the fact that theoretical considerations relied on spectral densities of irregularities and misalignments of the guide strings reported in [6], and derived from normative reference documents [9, 10] having relevance to conveyances in hoisting installations. Predicted data given in this study, based on irregularities and misalignments readouts in real conditions [11], are decidedly higher, yet the discrepancy between the theoretical and experimental results is still unacceptable.

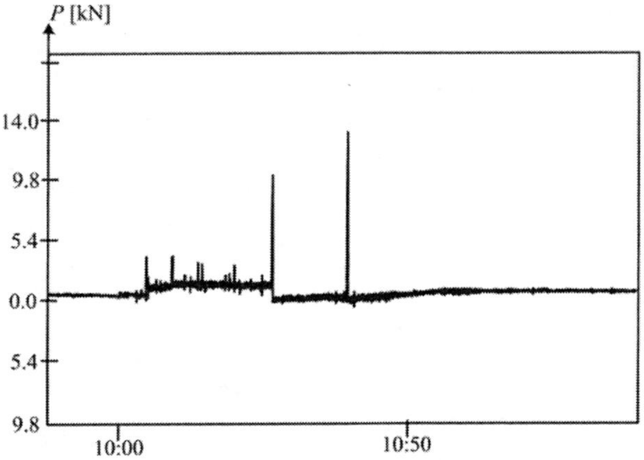

Fig. 9. Diagram of the conveyance -shaft steelwork interaction force [7].

Comparing and contrasting the experimental and predicted data will have its merits as long as all assumptions and conditions underpinning the dynamic analysis are fully satisfied. Actually, results of an experiment performed on the same object yet limited to measurements of the conveyance-shaft steelwork interaction forces summarised in [7] indicate the occurrence of impact effects in the systems whilst in fact the conveyance will run smoothly on the guide rails along short hoistway sections only. This behaviour is revealed in plots of conveyance-shaft steelwork interactions shown in Fig. 9 [7], where the peaks represent forces arising due to impacts between the skip and shaft steelwork.

It appears that when hoist operators attempt to distance the guides from the guide shoes, they do so for their own reasons. In order that theoretical results could be reliably verified experimentally[12], the conveyance needs to remain in contact with the guides, which is a requirement stipulated in relevant mining regulations.

References

1. S. Wolny, *Interactions in Mechanical Systems Due to Random Inputs on the example of a Mine Hoist*, International Education and Research Journal, vol. **1**. iss.5,s.72-76 (2015)
2. S. Wolny, *Displacements in Mechanical System Due to Random Inputs in a Mine Hoist Installation*, Engineering Transactions. Engng. Trans. **65**,3,p. 513-522 (2017)
3. S. Wolny, *Loads experienced by load-bearing components of mine hoist installations due to random irregularities and misalignments of the guide string*, Journal of Machine Construction and Maintenance, vol. **110**, iss. 3,p.79-86 (2018)
4. S. Wolny et al., *Developing critiria for assessing the supporting structure skip hopper of hoist installation in the terms of extending the period of safe operation*, Scientific an reasearch work, Department of Strength of Materials and Structures, AGH, Cracow (2003)
5. S. Wolny et al., *Development of multi-option modernization of mining cage lift R-VII*, Scientific and research work, Department of Strength and Fatigue of Materials and Structures, AGH, Cracow (2011)
6. S. Wolny, *Experimental verification of the state of stress in local Bering cables in a hosting installation due to random misalignments and irregularities of the guide strings,* (Article submitted for publication in Archives of Mining Sciences) (2018)
7. D. Bańdo, *The impast of selected operating parameters on the strength of structural elements of the skip hopper,* PhD thesis, AGH, WIMiR, Depatrment of Strength of Materials and Structures, Cracow (2006)
8. R. Gutowski, W. A. Swietlicki, *Dynamics and vibrations of mechanical systems*, PWN, Warsaw (1986)
9. S. Kawulok, *The influence of the shaft steelwork on mechanics of guiding skip hopper* , GIG works, Katowice (1989)
10. Mining Regulation: The Council of Ministers regulation of 30 April 2004 in the admission of using in mining plants (Dz.U. No. 99,item 1003 of 2005 No. 80, item 695 and of 2007 No. 249, item 1853, p.1.2 skip hopper).
11. F. Matachowski, *Developing design criteria of selected skip hopper elements,* PhD thesis, AGH, WIMiR, KWZMiK, Cracow (2011)
12. M. Szot, *Influence of irregularities of stiff guidance of shaft conveyances on the change of hoisting rope loading- methodology of testing, measuring equipment*, Mining and Environment, Quarterly 4, Central Mining Institute, Katowice (2003)

Anomaly detection in the concrete arc girder subjected to fatigue test

Dominika Ziaja[1,*] and *Piotr* Nazarko[1,**]

[1]Rzeszow University of Technology, Powstańców Warszawy 12, 35-959 Rzeszów, Poland

Abstract. A steel reinforced concrete arch of a bridge girder has been subjected to static and fatigue tests. The aim of this study is the application of guided waves in non-destructive diagnostics of civil engineering structures and early damage detection. Two piezoelectric transducers were mounted at a distance of 1 m to monitor area of the arch keystone. After every 500 000 cycles the signals of elastic waves have been measured and the girder visual examination was carried out. It turned out that both the load magnitude and the appearance of cracks have affected the signal changes. The obtained signal database has been used to train artificial neural networks and establish a diagnostic system. The results of the conducted tests have showed good sensitivity of anomaly detection and satisfying accuracy of load identification.

1 Introduction

The main problems of the bridge construction is a significant reduction in the durability of structures caused by corrosion of steel and concrete. One of the ways to improve the structures safety is to increase the inspections availability and its frequency by the use of structural health monitoring (SHM) systems. They are based on information obtained during non-destructive test. In case of crack detection in concrete structures the impedance-based damage detection method [1, 2], ultrasonic testing [3, 4] or Digital Image Correlation (DIC) [5] are used for this purpose. Also, elastic waves propagation phenomenon is widely applied [6, 7] and this approach has been investigated in the presented study.

The fundamentals and possible applications of guided waves is well described in the literature [8, 9], especially with respect to simple structures (i.e. beam or plate like structure, made of materials in homogeneous layers) with known boundary conditions. In general, a guided wave excited in the structure, propagates through the structural elements and changes its character (direction, amplitude, etc.) in case of damage appearance and growth (i.e. cracks, delaminations, corrosion, impact damage, yielding zone). This technique has gained considerable popularity in the monitoring of aircraft structures. Nowadays, it is also developed in terms of its applications in the diagnostics of building structures [10–12].

Tracking changes of the propagation character may allow to determine the location and size of damage. However, due to the large number of data, this process requires an adequate signals processing and reasoning algorithm. The dimensionality reduction of measurement data often requires the use of signal compression techniques such as wavelet decomposition,

*e-mail: dziaja@prz.edu.pl
**e-mail: pnazarko@prz.edu.pl

Hilbert-Huang transformation or independent component analysis. The approach used in this work is based on a simple transformation which converts the measurement data set into the space of principal components (PCA). It is a kind of lossy compression, but it usually works well in damage detection tasks [13].

An equally important issue is the correct design of the diagnostic system, sensitive to changes associated with the possible damage appearance and its growth. In this case, the inference is usually supported by expert systems, probabilistic methods and artificial intelligence algorithms. In this approach, artificial neural networks were used for this purpose. With their help, an attempt was made to carry out two separate tasks:

- the first was related to the prediction of load force applied to the tested girder – standard neural network (SNN) for regression task was used for this purpose,
- the second one concerned the classification of structure state on the basis of signals registered after each series of fatigue tests – autoasociative neural network (ANN) and novelty index (NI) were adopted in this case.

The application of neural networks allowed the automation of inference process what is a great advantage of the developed diagnostic system due to possible application in the continuous monitoring task.

The second chapter contains a description of the investigated object and the adopted research scope. The third chapter presents the idea of the developed diagnostic system, supplemented with exemplary results of its operation.

2 Laboratory tests

The steel reinforced concrete arch of a bridge girder (Fig. 1) has been subjected to static and fatigue tests. Its length and height were 9.95 and 2.29 m respectively, while the dimensions of the cross-section were 1.00 by 0.28 m. The research programme assumed 2 million cyclic loads. After every 500 000 cycles one single static test (up to 120 kN) and visual examination of the girder was carried out. As a result, the appearing and progressing concrete cracks were marked. In addition to continuous measurements of loads, deformation and stresses, two piezoelectric transducers (Noliac CMAP6, $5 \times 5 \times 2$ mm) were mounted at a distance of 1 m on the arch longitudinal axis and around the keystone (at the bottom side, see Fig. 1). They were used in order to excite propagation of elastic wave and to receive the structure response.

Figure 1. Laboratory model of the concrete bridge girder: general view and location of piezoelectric transducers (PZT)

During the static tests (an exemplary load history is shown in Fig. 2) the elastic wave signals were stored every 200 μs. As the excitation pattern 3.5 sine wave, modulated by Hanning window, was used. Its operating frequency was set to 56 kHz and the amplitude to 50 V (peak-to-peak). This parameters were tuned in order to obtain the highest possible signals to noise ratio.

The received structure response was passing through the charge amplifier (Model 463A, PCB) and then was recorded on a digital oscilloscope (LeCroy WaveRunner) with a sampling frequency of 5 MHz. Two examples of registered time signals of elastic wave (under load of 5 kN and 120 kN) were shown for comparison in Fig. 3. Clear differences between these two signals can be observed. Variance of all collected data in one single static test is shown in Fig. 4. This allows assuming that signal changes are large enough to identify the load or/and deformation magnitude on this basis. It is also worth mentioning (although such a comparison has not been provided in this article) that clear differences in signals were also visible after each of the series of fatigue tests when compared to the same load levels.

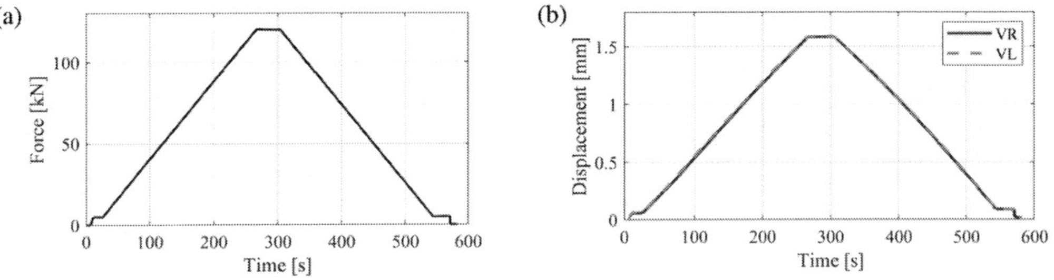

Figure 2. (a) The applied load history and (b) the vertical displacement recorded during static test (VR/VL – right/left side inductive sensor)

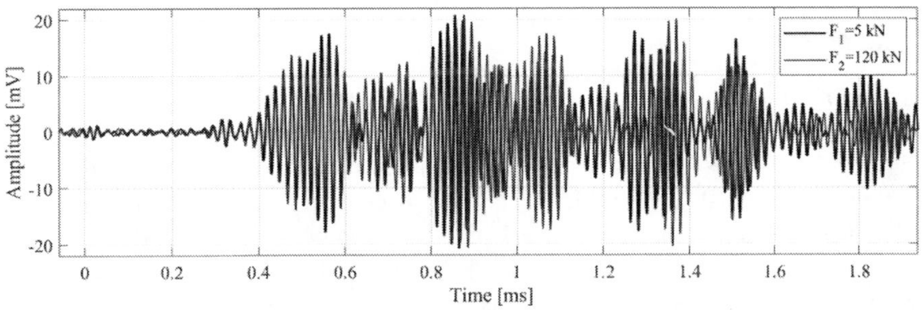

Figure 3. Exemplary time signals of elastic waves recorded in the concrete arch girder subjected to load of 5 and 120 kN

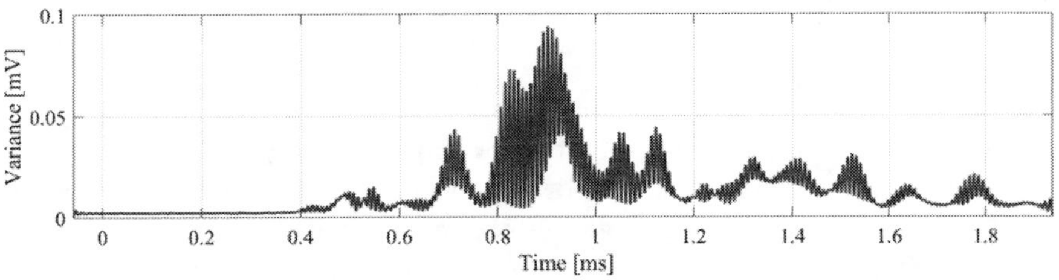

Figure 4. Variance of the elastic waves signals recorded in the concrete arch girder subjected to single static test up to 120 kN

3 Training the diagnostic system

3.1 Data processing

On the level of signal processing, the structure responses were filtered using bandpass filter with cut off frequencies set to 32 and 80 kHz. Then signal features extraction was performed using PCA, as described in [14]. Input vectors have been created based on the selected number of principal components obtained from the measured time signals. Projections of three first principal components related to elastic wave signals recorded during one single static test and after each series cyclic loads are shown in Fig. 5. Full-length signals were used for this purpose (2 ms). However, the conducted research shows [13] that PCA may be preceded by signals decimation procedure or their analysis may be limited to areas with the highest variability (see Fig. 4). This reduces the size of a coefficient matrix and speeds up the calculations related to PCA.

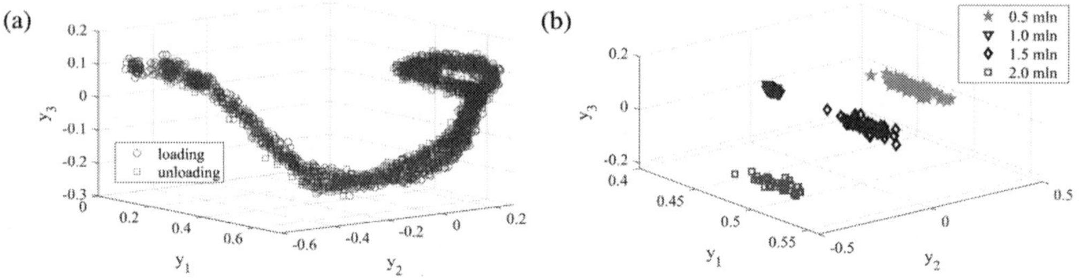

Figure 5. (a) Projection of three first principal components related to elastic wave signals recorded during single static test up to 120 kN and (b) after each series of cyclic loads (comparison for one load level of 5 kN)

3.2 Load prediction

Principal components related to single static test performed after 0.5 million cycles (Fig. 5) have defined a first pattern database (2909 patterns were stored). Then, it was divided into learning (1939) and testing sets (970). Every fifth pattern was used for testing (constant distribution). In this task SNN with one hidden layer were used. The size of the input layer was related to the number of considered principal components and was set in this case to 16. The output vector consisted of the load magnitude (see Fig. 2a). Exemplary results of the load prediction using SNN with architecture 16-5-1 are shown in Fig. 6 for both learning and testing patterns. It can be noticed there that very good agreements of the output and target values were achieved. The obtained statistical parameters of the testing error were the following: mean 1.21 kN, standard deviation 0.99 kN, maximum 6.03 kN.

3.3 Anomaly detection

This time the patterns database consisted of the principal components related to the load level of 5 kN selected from the static test conducted between every 0.5 million load cycles. It was assumed that each time the structure will be loaded with the same weight, i.e. 20-ton truck standing on or crossing the bridge. In this way 231 patterns were collected, taking into account only first part of the loading stage (see Fig. 2a).

The defined pattern database (Fig. 5b) was then used to train ANNs for the purpose of anomaly detection. An output vector of this neural network is the same as the input vector. Thus, when such a trained ANN is fed with the inputs obtained from a damage state of the

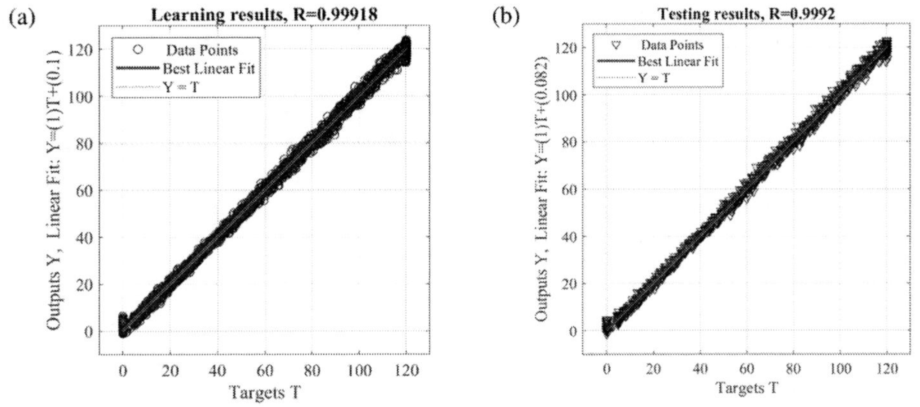

Figure 6. Exemplary results of the load prediction using neural network: a) learning, b) testing

system, the novelty index NI $= ||x - x||$, which measures the distance between the known input and output of the ANN, will increase [14, 15]. In this study, feedforward neural networks with one hidden layer were used and they were trained using Levenberg-Marquardt algorithm.

The learning set (76 patterns) was related only to the signals measured after the first series fatigue tests. Every fifth pattern was then selected to the validation set (15 patterns). All other patterns defined the testing set. After ANN training the NI was calculated for every single pattern. Its exemplary values, obtained using ANN's architecture 5-3-5, were shown in Fig. 7.

Introducing certain threshold levels enables not only the anomaly detection but also patterns separation and its classification. The threshold level for the reference set may be calculated as follows: $t = \frac{1}{n}\sum_{i=1}^{n} \mathrm{NI}_i + 5\sigma$, where σ is a standard deviation calculated for NI_i and n is a number of reference patterns used for training the diagnostic system. Its value was marked in Fig. 7 with a dashed line.

If we hypothetically assume that after a series of 1 million cycles a visual inspection of the structure was carried out and any serious defects were found (i.e. the limit value of the crack width was not exceeded), the recorded signals can be included into the reference set. Exemplary results of anomaly detection obtained in this way were shown in Fig. 8. It can be seen there that also this time the obtained results enable flawless separation of the analysed patterns classes.

4 Conclusions

On the basis of experimentally measured elastic wave signals, two diagnostics tasks have been accomplished: 1) prediction of load magnitude, 2) anomaly detection after a series of fatigue

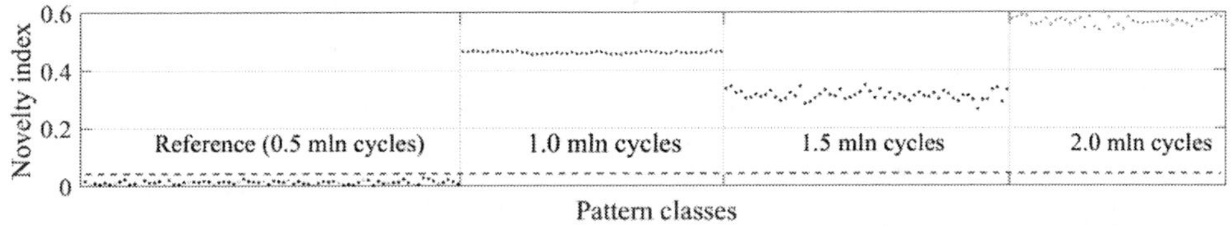

Figure 7. Novelty index calculated on the basis of exemplary results of the ANN training (5-3-5) – the learning set consisted of the patterns related only to the signals measured after the first 0.5 million of cycles

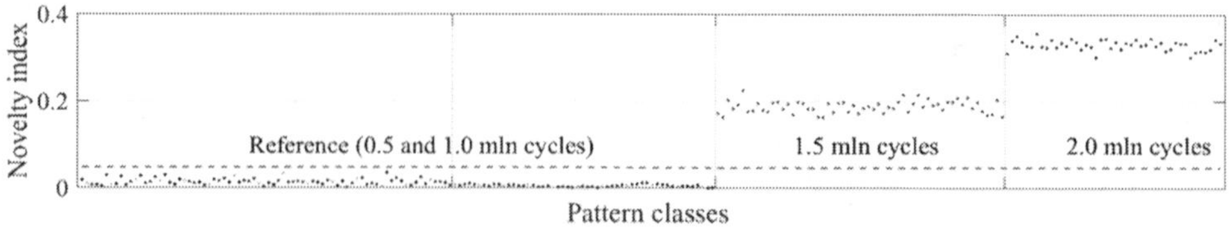

Figure 8. Novelty index calculated on the basis of exemplary results of the ANN training (5-3-5) – the learning set consisted of patterns related to the signals measured after 0.5 and 1.0 million of cycles

tests. The obtained results showed good accuracy and sensitivity of anomaly detection. It was possible thanks to the application of ANN and SNN. The discussed approach enables also the automation of the diagnostic process, which increases both the availability of inspections and the structure's safety.

Unfortunately, at the current stage of research it was not possible to determine how small changes can be detected, because the visual assessment of its condition was carried out periodically after the end of each series of fatigue loads. It can only be concluded that each series of fatigue cycles introduced qualitative changes in the state of the analysed areas, without determining whether they are relevant from the point of view of its safety.

During future tests, it is worth checking if the trained SNN is able to predict load magnitude also in case of the larger number of cycles (e.g. 1 million). In case of anomaly detection the patterns database can be extended with signals measured during unloading stage. An alternative patterns database can be also created when the other load levels are considered, e.g. 120 kN. The research plan assumes that in the near future similar tests will be carried out with respect to the concrete arc reinforced using composite fibre bars.

References

[1] S. Park, S. Ahmad, C.-B. Yun, Y. Roh, Experimental Mechanics, 46 (5), 609–618 (2006)
[2] R. Silva, K. Tsuruta, D. Rabelo, R. Finzi Neto, V. Steffen, EWSHM (2016)
[3] A.A. Shah, Y. Ribakov, Materials & Design, 30 (9), 3504–3511 (2009)
[4] S. A. Kumar, M. Santhanam, Proc. National Seminar on Non-Destructive Evaluation, Dec. 7–9, 2006, Hyderabad
[5] T. M. Fayyad, J. M. Lees, Procedia Materials Science, 3, 1585–1590 (2014)
[6] G. Zhao, D. Zhang, L. Zhang, B. Wang, Sensors, 18, 4176 (2018)
[7] Y. Lu, J. Li, L. Ye, D. Wang, Constr. Build. Mater., 47, 370–378 (2013) [
[8] T. Stepinski, T. Uhl, W. Staszewski, *Advanced Structural Damage Detection. From Theory to Engineering Applications* (John Wiley & Sons, Ltd, 2013)
[9] Z. Su, L. Ye, *Identification of Damage Using Lamb Waves. From Fundamentals to Applications* (Springer, 2009)
[10] P. Nazarko, MATEC Web of Conferences, **262**, 10009 (2019)
[11] M. Rucka, B. Zima, R. Kędra, Archives of Civil Engineering, **60**, 493–515 (2014)
[12] J. Li, Y. Lu, R. Guan, W. Qu, Construction and Building Materials, **131**, 388–399 (2017)
[13] P. Nazarko, L. Ziemianski, Engineering Failure Analysis, **69**, 97–107 (2016)
[14] P. Nazarko, Inverse Problems in Science and Engineering, **21** (6), 945–956 (2013)
[15] M. R. Hernandez-Garcia, M. Sanchez-Silva, Computational Intelligence and Its Applications, Ch. VIII, 158–187 (Idea Group Publishing, 2007)

AUTHOR INDEX

Adrian, Pawelek 89
Agata, Drzewosz 177
Andrzej, Pochanke 129
Arkadiusz, Tilsen 161
Barbara, Markiewicz 67
Bartlomiej, Dyniewicz 15
Bartosz, Miller 83
Czeslaw, Bajer 15
Damian, Gasiorek 9
Daniel, Szynal 137
Dawid, Zwolinski 105
Dominika, Ziaja 67, 185
Ewa, Blazik-Borowa 1
Filip, Matachowski 75
Grzegorz, Piatkowski 67
Jacek, Chróścielewski 153
Jaroslaw, Bec 1
Krzysztof, Kozioł 37, 43
Krzysztof, Wilde 161
Larin, Denis V. 121
Leonard, Ziemianski 83
Lucjan, Janas 137
Lukasz, Kutrowski 145
Lukasz, Pyrzowski 153
Lukasz, Szyszka 67
Mariusz, Pawlak 97
Mateusz, Rajchel 113
Piotr, Fedelinski 23
Piotr, Lichota 89
Piotr, Nazarko 185
Przemyslaw, Wielentejczyk 59
Robert, Konowrocki 129
Roman, Lewandowski 59
Ryszard, Holubowski 31
Savin, Sergey N. 121
Sebastian, Slawski 97
Sebastian, Uzny 145
Slawomir, Badura 75
Slawomir, Duda 97
Smirnova, Elena E. 121
Stanislaw, Burzynski 161
Stanislaw, Wolny 169, 177
Tomasz, Ferenc 153
Tomasz, Siwowski 113
Tomasz, Szolc 129
Tomasz, Wiczenbach 153
Waldemar, Latas 51
Wojciech, Danek 9
Wojciech, Witkowski 161
Zdzislaw, Pawlak 105

EDP Sciences – Web of Conferences
17, Avenue du Hoggar
Parc d'Activité de Courtabouf
BP 112
F-91944 Les Ulis Cedex A
France

ISBN 978-1-5108-9070-1

CURRAN ASSOCIATES INC.
proceedings.com

3rd International Conference on Chemical and Biological Sciences 2016

MATEC Web of Conferences Volume 60 (2016)

Amsterdam, Netherlands
23 - 25 March 2016

edp sciences